# BIOLOGY

# THE AUTHORS

**Gordon Alexander** received his Ph.D. degree from Princeton University and joined the Department of Biology of the University of Colorado in 1931. He became a full professor in 1939 and was head of the department from 1939 to 1958. He is now Professor Emeritus. Dr. Alexander was Visiting Professor of Biology at Chulalongkorn University, Bangkok, Thailand, 1928–1930, and Fulbright Lecturer at the same institution 1956–1957.

He is a member of several scientific societies. He has been on the executive committee of the Ecological Society of America and on the editorial board of *Ecological Monographs;* he has represented the Entomological Society of America on the Council of the American Association for the Advancement of Science; he has been president of the Colorado-Wyoming Academy of Science.

Professor Alexander is the author of *General Zoology* (a companion Outline) and *General Biology* (a standard textbook), as well as numerous articles on the biology of Thailand, animal distribution in relation to altitude, and other aspects of biology. At present he is conducting research on animal distribution in relation to altitude.

**Douglas G. Alexander** received his Ph.D. degree from the University of North Carolina. He has been at Chico State College, California, since 1965, and is now Associate Professor of Biology. He has also taught at the University of North Carolina and the University of Colorado. His earlier biological experience included work with the U.S. Fish and Wildlife Service and, in marine ecology, at the Marine Biological Laboratory, Woods Hole, Massachusetts. He was a member of a bird and mammal collecting expedition in northern Thailand in 1956.

He is a member of the Ecological Society of America, the American Society of Ichthyology and Herpetology, and other scientific organizations. His current research is on aquatic ecology in the central California valley.

# BIOLOGY

**GORDON ALEXANDER**
**DOUGLAS G. ALEXANDER**

Ninth Edition

BARNES & NOBLE BOOKS
A DIVISION OF HARPER & ROW, PUBLISHERS
New York, Hagerstown, San Francisco, London

# PREFACE

When the first edition of *Biology* appeared, in 1935, no one talked of "molecular biology," and physical scientists were still scornful of biology as "descriptive science." Biology has so changed since that time, however, that it has not only won the respect of physical scientists but is now recognized as a science that has much to offer the scientifically curious investigator.

In the period since 1935 many basic biological discoveries have been made. The chemical nature of viruses has been determined. The processes of respiration and photosynthesis have been analyzed into complex series of reactions. The chemical machinery by which energy is mobilized for use in living cells has been discovered. The structures and functions of minute cell constituents, such as mitochondria, are now better understood. The structures and roles of nucleic acids in development, differentiation, and inheritance are being unravelled. We even have the beginnings of a coherent theory to explain the origin of life on this planet.

These discoveries, properly in the field of molecular biology, have not been the only profitable ones. We now see living nature as part of a large-scale system, energy for its operations being derived by producer organisms from the sun and passed along to those organisms that benefit secondarily. This fundamental ecological principle and its corollaries are basic to man's analysis of his current problems of population and environment. As we have known for a long time, species within natural communities undergo changes. Now we see this evolution as involving the genetics of populations. While the causes of the great variety in nature still constitute the central problem in biology, the next few years will undoubtedly be marked by increased understanding of this problem.

Recent discoveries have made no change, of course, in the *purpose* of this Outline. As in previous editions, it is designed for use with many types of general biology courses. The sequence of material varies greatly among such courses, so we have adopted what seems to us a logical and usable sequence, one that is similar to that in many widely used textbooks. Related materials have been placed together, facilitating the use of *Biology* as a reference; and its scope is comprehensive enough to justify its use as a condensed textbook of biology.

To make fullest use of this Outline students should refer to the section on How to Study Biology immediately following the Table of Contents. Note that the Appendix includes, in addition to other useful reference material, a list of published References and a Glossary.

While this book will provide useful review material for students in courses in botany or zoology, those desiring Outlines particularly in these fields should refer to *General Botany* by Harry J. Fuller and Donald D. Ritchie or *General Zoology* by the senior author of the present work. Students who find it necessary to review basic concepts in chemistry should see *College Chemistry* by John R. Lewis.

G. A.

D. G. A.

# ACKNOWLEDGMENTS

The ideas incorporated in this book have had such a complex history that it is difficult now to trace the influences behind them and to make appropriate acknowledgments. We are indebted to many individuals, particularly to those teachers, colleagues, and students with whom we have been directly associated. We are especially indebted to members of the editorial staff of Barnes & Noble, in particular to Jeanne Flagg, biology editor, for her extensive helpful advice during the preparation of this completely reset revision. Specific acknowledgments for the use of copyrighted materials are made where appropriate. Most of the illustrations, however, are original with us.

We will appreciate suggestions and criticisms designed to improve the usefulness and accuracy of this Outline, and we will welcome these from students as well as instructors. We are grateful to many users of previous editions for contributions of this kind.

G. A.
D. G. A.

# CONTENTS

# *Contents*

## PART FIVE: RELATIONS OF BIOLOGY TO MANKIND

## APPENDIXES

# HOW TO STUDY BIOLOGY

Biology is the science of living things. It deals primarily, therefore, with objects about us, not human ideas. Such a science is not and cannot be based on "book learning," and those who study it today use books only to acquire information that was acquired originally by direct observation of nature. The following study suggestions are based therefore on the realization that lectures and textbooks are aids in the study of biology but that a real understanding of the science is acquired only when life is observed directly. The fundamentals of biology are best acquired by the direct process of observation by the student. Laboratory or field work is essential, and the student in a course without organized laboratory work should make up for this deficiency by a conscious effort to apply what he learns in class to the world of living things around him.

**Scheduling Your Time.** Course programs are set up to require approximately three hours per week of a student's time for every credit hour. Thus, a course that meets three lecture hours and one three-hour laboratory period each week, and that probably gives four credit hours, demands an average of six hours per week in outside study. Use such a figure in determining the time you should schedule for study, and distribute that time over several days. Concentrated study for periods of one or two hours is better than attempting to cram a week's study into one day.

**Learning the New Vocabulary.** Every science deals with exact expression—the right word, correctly spelled. In the biological sciences a specialized and technical vocabulary is essential for complete comprehension of the subject matter. The new vocabulary is large, but it is not difficult to acquire if each new term is made a part of your vocabulary when it is first introduced: its meaning (and its spelling) should be understood by use, not learned by rote. The best method of learning meaning and spelling together is to analyze the origin of the word. Most biological terms are of Greek or Latin origin, but that does not mean that one must study either language to learn the meanings of the roots of English words. Many commonly recurring roots are not difficult to recognize or to learn, and the derivations of most new terms can be found in a modern unabridged English dictionary. If you acquire the new vocabulary by learning its origins, you will soon find that you already know the roots for various new terms as they appear; acquiring the vocabulary therefore becomes simpler rather than more complex as time goes on.

**Getting the Most from Lectures.** Textbook assignments are made to correspond to lecture material. Therefore, you should read your assignments before going to class. Good lectures are not simply a rehash of textbook material, but, even if they were, one could get much more out of them after having done some reading on the subject ahead of time.

Useful lecture notes are very important. Record these legibly in permanent form (in ink) during the lecture. You should not have to waste time rewriting your notes afterward; instead, organize the material of the lecture at the time of the lecture. Even if the lecturer does not indicate the principal topics, these should be obvious if you have read the assignment.

Discipline yourself to take down only the important points. If you try to record every statement, you soon become so involved in the mechanics of the process that you listen for words instead of ideas.

**Getting the Most out of Textbooks.** The major problem in reading, assuming that the ordinary English vocabulary is understood, is concentration. Speed in reading the average textbook in biology is not essential. Rapid reading for a general impression is definitely disastrous. The average assignment in a biology textbook is short in number of pages but full of details that should be mastered. Therefore, you must read for retention of details as well as for an understanding of principles. On the other hand, memorizing facts without understanding them is of no value, so read for comprehension as well as retention.

A good environment during study time is necessary. Have a comfortable chair (but not one best used for lounging), an adequate desk or table, and good light. Distractions (human voices being the most serious) should be eliminated as far as possible. It may be better to study in a library reading room than in a dormitory.

Outlining the textbook is a good idea if your outline is not merely a copy of subdivisions already indicated in the text. Underlining key words and phrases, and using special notations in the margins, may prove as useful as a more formal outline. Nothing is gained by underlining several consecutive lines of type, however; that simply makes the passage more difficult to read.

**Getting the Most out of Laboratory Work.** Laboratory studies provide the firsthand experience so necessary in acquiring familiarity with the materials and methods of science. Books cannot provide this. You must therefore develop the habit of asking questions of the laboratory material (not your instructor), working and thinking independently. Only in this way can you appreciate the objective point of view and acquire the critical approach essential in the scientific method. You will probably be asked to perform experiments as well as make simple observations. All such assignments are designed to help you understand biological principles, so approach them with the question "Why am I doing this?" on an intelligent rather than superficial basis. Always read the directions carefully before beginning an experiment or series of observations, and then follow the suggested sequence.

More than likely, your laboratory work will involve considerable use of living organisms. When using living plants or animals follow the special instructions for their care. Such instructions will vary considerably from one type of organism to another. One should remember, for example, that living organisms being examined under a microscope must be kept continually moist. Preserved specimens that have been in liquid should also be kept moist while being studied. Water may be used for this purpose in most cases, but if the specimens are to be used again they must be stored in appropriate containers, completely submerged in the preserving fluid.

Prepared microscope slides are easily broken and should be handled with care. They should be cleaned and dried before being put away, care being exercised to avoid pressure on the cover glass. When you focus a slide under the microscope, bring it into focus under low power and center the objects you wish to study before transferring to high-power magnification. (Some slide preparations are too thick to be examined under high power so first determine if the high-power lens can be used.) And note especially: Never carry a microscope when it has a loose slide on the stage.

Keep your dissecting instruments sharp. They should be cleaned and dried carefully each time before they are put away. The acquisition of skill in their use should be one of your aims. This means using scissors and forceps much more frequently than they are used by most students.

Records of laboratory work required by an instructor generally take two forms, drawings and written reports. Both types of records have the same three functions: (1) They help fix the observations in your mind in a way that simply making the observations will not do. (2) They give you a record to use when you review the work. (3) They indicate to your instructor that you have completed the assignment.

A drawing made from your own observations will teach you much more than one made by somebody else, no matter how good it is. And everybody can learn to draw, at least well enough for this purpose, for the methods of the mechanical draftsman are more appropriate in the biology laboratory than are those of a freehand artist. The biologist's aim is to prepare an accurate "rendering," not a work of art. To assure this accuracy you must take a series of critical measurements and transfer these to the drawing paper, then fill in the details freehand.

Drawings made of objects under a microscope present special problems. You can learn to transfer such images to the drawing paper, however, when you become accustomed to using both eyes at once. (By this same method you can learn to calculate magnification by the microscope.) Look through the microscope at the focused object with the left eye but keep the right eye open. Now bring a sheet of drawing paper close to the base of the microscope on the right side. With a little practice you will soon be able to see the object under the microscope superimposed on the drawing paper. You can even learn to see a six-inch ruler placed on the drawing paper, and you can use it for measuring the magnified image. The process of drawing then becomes little more than the process of "tracing" the object as seen in the field of the microscope.

Standard outlines (e.g., Purpose, Materials and Methods, Results, Conclusions; or Materials, Procedure, Questions) are often used in write-ups of laboratory experiments. If essay-type reports are called for these should consist of complete sentences, in appropriate paragraph organization, grammatically correct, and with (as in the case of labels on drawings) letter-perfect spelling. If you are to describe the materials and methods be sure you understand the operation of any special apparatus used. Also, be sure to list the kind or kinds of organisms or other materials used. Remember that, basically, the purpose of an experiment is to test some variable. Be aware of that variable, and report your conclusions in terms of that variable.

**Supplementary Study.** Much can be gained by examining library sources in addition to standard textbooks. Assuming that the books are on open shelves and under the widely used Dewey Decimal System of library classification, you will find the most useful references under 570 (Biology), 580 (Botany), 590 (Zoology), 612 (Medicine), and 630 (Agriculture). If the Library of Congress System of classification is used, the most valuable references will be under QH (Biology), QK (Botany), QL (Zoology), QP (Physiology), R (Medicine), and S (Agriculture). Subject catalogues may be consulted if one may not browse along the shelves.

Museums, like libraries, are valuable sources of supplementary study. Some university and college museums have special sections devoted to surveys of the plant and animal kingdoms. Such sections are particularly useful in reviewing the characteristics of different groups.

Man's environment is full of living organisms, plants and animals. As a student of biology you can gain a great deal by carrying over into everyday life what you learn of animals or plants in the formal course. There are actually many opportunities to do so—in the parks, in markets, on the farm, or along the beach. Use your increasing competence to understand better and to derive educational benefit from daily contacts with biological information. Many new advances and basic concepts relating to biology are included in television and radio features as well as in newsstand reading. Your own training should make you a good judge of the significance and worth of the many public communications we now receive on public and personal health, population problems, pollution, and conservation.

**Reviewing the Course.** An adequate review of a course in biology involves going over the textbook, lecture notes, and laboratory notes. Recognize the major generalizations of the course but don't forget the details that justify the generalizations. In reviewing the characteristics of different phyla or lower groups of animals or plants, try to follow a definite pattern or outline. Such an outline, applicable to all groups, might be the following: Metabolism, Irritability, Reproduction (each of these headings suggesting both structures and functions), Distribution, Evolutionary Relationships.

**Writing the Final Examination.** Examinations have several purposes. For students the most important are two: They stimulate review, which makes for better retention. And they test recall of facts and comprehension of principles. In answering the examination take nothing for granted. Assume that you are writing the examination for someone who knows none of the answers. The instructor is not expected to read between the lines. Your answers must therefore be complete and clear.

Planning for an examination should come early. Keep three types of review in mind, text and outside reading assignments, lecture notes, and laboratory records. If you keep up with your work in all three throughout the term, an examination should have no terrors. A review of a few hours, scheduled before the last night, should then prove adequate. Cramming may be better than nothing, but it has little to recommend it. Relax the night before an examination and get a good night's sleep.

For examples of examination questions, as well as additional suggestions on writing final examinations, see Appendix E, at the end of this book.

# BIOLOGY

*Chapter I*

# INTRODUCTION

Biology is the science of life in all its phases. The term (from *bios,* life and *logos,* discourse) was first used in 1802, when it was introduced independently by Lamarck and Treviranus. Biology includes botany, the study of plants, and zoology, the study of animals.

## RELATION OF BIOLOGY TO OTHER SCIENCES

Basic to the study of biology are mathematics, physics, and chemistry. Biology contributes to and receives information from geology, sociology, anthropology, and psychology, the last three being in part specialized fields of biology. It underlies the applied sciences of medicine and agriculture, the latter term being used in its broadest sense to include forestry, fish culture, etc.

## SCIENTIFIC METHOD

One of the characteristics of a science is that the knowledge comprising it is acquired by the scientific method. This is the application of a critical, unprejudiced analysis to the determination of causal relations in nature. It may involve *observation* under natural conditions alone, or it may involve observation with control of variable factors by the scientist, i.e., *experiment.*

The scientific method is usually applied in a series of steps:

(1) *Observation* of a phenomenon and the recognition of a problem.

(2) The construction of a *hypothesis,* or the attempt to explain a phenomenon.

(3) *Experimentation,* or the testing of the hypothesis.

(4) *Conclusion,* if the test demonstrates the correctness of the hy-

pothesis. If the hypothesis proves to be incorrect a new one must be set up and tested. The formulation of a *theory* * or generalization follows if repeated experimentation demonstrates that the hypothesis is valid.

Theories are arrived at by *inductive reasoning*, which is reasoning from particular observations to generalizations. Once a theory is formulated it can be applied to new situations; it can also be used to predict certain results. This type of reasoning, applying generalizations to particular situations, is *deductive reasoning*. Both types of reasoning occur in scientific thought.

With the discovery of new facts and relationships, usually the result of improved technology, a theory may have to be revised or discarded. Those theories that have been found to have universal validity and predictability may become known as *principles* or *laws*.

## SUBDIVISIONS OF THE SCIENCE OF BIOLOGY

Much of present-day biology is called *molecular biology*. This is not significantly different from biochemistry except that the molecular biologist is concerned primarily with the large molecules that characterize living organisms. The subject matter of molecular biology will be summarized in Chapters II, IV, and V. The biology of organisms, sometimes called supramolecular biology in contrast to molecular biology, is divided into subsciences on the basis of (1) the kinds of organisms studied or (2) their levels of biological organization.

**Subdivisions Based on Kinds of Organisms.** The "classic" division of biology is into two subsciences, the study of plants (*botany*) and the study of animals (*zoology*). A third subscience, *microbiology*, the study of microscopic organisms, is often recognized. Differences between these subsciences are not always clear, however, for there are organisms that are on the border line between plants and animals, and organisms within the province of microbiology that are transitional between microscopic and larger organisms.

BOTANY. The science of plant life is botany. It has many subdivisions, such as *agrostology*, the study of grasses; *bacteriology*, the study of

---

* In common usage, the word theory usually refers to a generalization that has not been tested, in other words to what a scientist calls a hypothesis.

bacteria, considered also a part of microbiology; *dendrology,* the study of trees; and *mycology,* the study of fungi, considered in part under microbiology.

ZOOLOGY. The science of animal life is zoology. Like botany it has many subdivisions, such as *entomology,* the study of insects; *parasitology,* the study of animal parasites, in part under microbiology; *ornithology,* the study of birds; and *protozoology,* the study of unicellular animals, considered also a part of microbiology.

MICROBIOLOGY. This is the science of unicellular organisms, viruses, and related forms. It includes the subject matter of bacteriology and protozoology, and, in addition, it deals with viruses, unicellular plants other than bacteria, and microscopic organisms of uncertain status.

**Subdivisions Based on Levels of Biological Organization.** Biology can also be subdivided on the basis of levels of organization of living systems: cells, organisms, populations, and communities. Although this method of subdivision avoids the dilemma of trying to distinguish between plants and animals, it does not completely eliminate overlapping subject matter. (Subjects like embryology and genetics, for example, involve several levels of organization.)

THE BIOLOGY OF CELLS. The study of cells is *cytology.* This can be divided into *morphological cytology,* the study of cell structure, and *cell physiology,* the study of cell functions. These divisions are somewhat arbitrary. An understanding of cytology is basic to an appreciation of physiology, growth, differentiation, and inheritance in all types of organisms.

THE BIOLOGY OF ORGANISMS. Subsciences that deal with individual organisms and their larger components are: *morphology,* the study of structure; *physiology,* the study of function; and *embryology,* the study of development. The morphology of multicellular animals includes *anatomy,* the study of gross structures (e.g., organs), and *histology,* the study of tissues (groups of cells of similar structure and function). The study of tissues in plants is generally referred to as *plant anatomy.*

THE BIOLOGY OF POPULATIONS. Under this level are subsciences in which organisms of the same species are studied in group relations. *Taxonomy* is the science of naming and classifying the different kinds of organisms. *Ecology* is the study of the relations between organisms and environment, involving an explanation of why a population is found where it is and in its given abundance. *Phytogeography* and *zoogeography* are the subsciences that deal, respectively, with plant and animal distribution in the larger geographical divisions of the earth.

*Genetics* is the study of biological inheritance, the degrees of similarity and difference between parents and offspring, and the factors that control these similarities and differences. Because it involves the study of the transfer of inherited information from cell to cell, and thereby from generation to generation, genetics also bridges cellular and organismal levels of organization. *Evolution* (*phylogeny*) involves the study of existing adaptations of organisms and the changes through time that have produced these characteristics. Organic evolution is the term applied to the progressive development of more complex forms of life from simpler ones.

THE BIOLOGY OF COMMUNITIES. Studies of the natural relationships of groups of organisms of different species constitute these subdivisions of biology. The study of communities is, of course, a part of ecology. The ecological study of the dynamic interactions within a community of organisms and its nonliving environment is called *ecosystem ecology*. The analysis of pollution and problems of conservation are considered within this subscience.

# LIFE—ITS CHARACTERISTICS

Although one can readily distinguish between the living and the nonliving state in familiar organisms, the word *life* cannot easily be defined. It is best described in terms of its special attributes, as follows:

**Organization.** Living matter is organized in a series of levels of increasing complexity: cells (Chap. III), organisms (Chap. VII), populations (Chap. XX), and communities (Chap. XX). Living matter itself is commonly called *protoplasm* (Chap. II), but protoplasm is highly variable and the term does not imply any definite kind of organization. The simplest form of independent biological organization is the cell. A cell does, however, have subcellular units called *organelles,* which have characteristic structure and function. Three organizational categories occur between the levels of cells and organisms: *tissues,* which are groups of cells with the same structure and function; *organs,* which are groups of cells or tissues associated with particular body functions; and *organ systems,* which are groups of functionally related organs.

**Metabolism.** Metabolism is the sum total of all chemical processes going on in living matter. These are of two general types, synthetic or constructive reactions, in which energy is stored in chemical compounds, and analytic or destructive reactions, in which energy is released. The former type, including food manufacture and the condensation of simple foods into larger molecules, is called *anabolism*. The latter type, including the decomposition (digestion) of large molecules and the oxidation of the smaller compounds derived from them, is called *catabolism*. These reactions, which are characteristic of living matter, are made possible by the existence of specialized organic catalysts, *enzymes*. (See Chap. IV.)

**Irritability.** Living matter responds to a variety of external stimuli, or changes in the environment. Stimuli produce changes in cellular matter that may initiate responses in the organism. These responses are induced by *impulses* or by secretions known as *hormones*. In unicellular organisms, impulses spread to other parts of the cell from the place of stimulus; in multicellular forms, impulses or hormones are conducted to other parts of the organism. These may cause special responses, e.g., contraction of a cell, cell growth, or secretion of a fluid. *Reception* of the stimulus, *conduction* of an impulse, and *response* to that impulse are all different aspects of irritability.

**Growth.** An excess of synthetic over destructive processes in metabolism results in an increase in size. This process of growth does not consist of the addition of material on the surface, as in the growth of a crystal, but of an increase in all parts, a growth involving an increase in the number of molecules present.

**Reproduction.** When a cell has grown to a characteristic size, it divides, forming two cells. In a unicellular organism this is reproduction; in a multicellular organism it is reproduction of cells, which, in such organisms, is an aspect of growth rather than reproduction. Reproduction in multicellular organisms may take place *asexually* (only one parent being involved) or *sexually* (two parents involved). In either case, offspring resembling the parent or parents are produced.

**Variation.** Reproduction of organisms involves some variation. Thus, although the offspring are similar to the parents they also present variations, either through *new combinations* of characteristics or through the chance occurrence of divergent characteristics (*mutations*).

**Adaptation.** The innate fitness of an organism for the environment in which it lives and thrives is called adaptation. Adaptation develops over a period of time through the interplay of inherited variations in successive generations and the fitness for a particular environment that these variations impart to the organism.

# THE ORIGIN OF LIFE

There are two theories of the origin of life on our planet. One of these is that life in simple form came to the earth from another planet or planetary system. Few scientists accept this theory, however, because, even if a living particle could travel through space, there is doubt that it could survive the extreme cold and the exposure to lethal radiation. Also, there would still be the question of how this living material arose at its place of origin.

The alternative theory is that life evolved on the earth from nonliving matter. This could have happened about two billion or more years ago, when the cooling planet had reached a stage when complex chemical compounds could have formed spontaneously. The steps between nonliving and living matter were probably gradual but must have involved, sooner or later, the formation of macromolecules, including enzymes, and, eventually, self-replicating nucleic acids. (See Chaps. II and IV.) These macromolecules could eventually have formed operating units, precursors of cells, that grew and reproduced through the consumption of environmental resources.

Evidence for the theory that life arose on the earth from nonliving matter comes in part from the fact that all chemical elements that occur in living matter are also constituents of nonliving matter. Furthermore, the basic uniformity of organic material implies a common origin. Other evidence comes from the fact that organic compounds have been formed under laboratory conditions that were attempted replicas of our planet's early environment. However, this explanation of the origin of life remains a theory, and the search for evidence will continue.

# BASIC CONCEPTS OF BIOLOGY

The most important principles derived specifically from biology are the following:

The Cell Concept (Chap. III)
The Organismal Concept (Chap. VII)
The Concept of Biological Inheritance through Chromosomes (Chaps. V & XVII)
The Concept of Organic Evolution (Chap. XVIII)
The Ecosystem Concept (Chap. XX)

*Part One*

# LIFE IN ITS SIMPLEST FORMS

# Chapter II

# THE PHYSICOCHEMICAL BASIS OF LIFE

Living material has certain characteristics that distinguish it from non-living material (Chap. I). These attributes result from a distinctive combination of physical and chemical properties. The mixture that displays these properties has been called *protoplasm.* The term was introduced by the Bohemian zoologist Purkinje in 1839 and was soon thereafter brought into general use among scientists by the German botanist von Mohl. Protoplasm was called the "physical basis of life" by T. H. Huxley, in 1869, and that is still a good definition for it.

Some early biologists conceived of life as resulting from unique physical or chemical properties of protoplasm. However, with the development of biochemistry as a science, it became evident that no one compound or no one physical state (such as a colloidal system) confers the characteristics of life upon matter. Some biologists have recently favored eliminating the word protoplasm on the ground that it implies a specific physical or chemical homogeneity. However, the term is still useful in referring to the living matter of the cell. Since biologists have for many years used the word as a general term, without implying homogeneity, we favor retaining it.

The structural unit of living matter is the cell. Although cells are diverse in size and form, they display the characteristics of life, and in so doing they have basic similarities, providing a spatial segregation of discrete structural units and a corresponding localization of chemical activities. However, some of the smallest units that possess the characteristics of life are not typical cells. Furthermore, free-living unicellular organisms and the tissue cells of higher organisms are modified in distinctive ways. Thus there is a certain ambiguity about the word "cell." Variations on this structural pattern will be considered in Chapters VI and VII. Before discussing cell structure (Chap. III) and cell function (Chap. IV) we shall consider the physical properties of living matter and the types of chemical compounds found in it.

## THE PHYSICOCHEMICAL BACKGROUND

Certain principles basic to an understanding of physics and chemistry are also important in biology. Living matter, like nonliving matter, is composed of *elements,* which unite to form *compounds.* The energy that goes into the formation of a compound is stored as *potential energy;* it is released as *kinetic energy* when the compound breaks down. In both living and nonliving systems, energy may change from one form to another, e.g., from potential energy to the kinetic energy of heat or movement, and from light energy to chemical (potential) energy. Transformation of chemical energy into various forms of biological activity is accompanied by a change of part of the energy into heat, which is dissipated. Thus such transfers are not 100 percent efficient. While no energy is lost (first law of thermodynamics), there is a decrease in usable energy in the system (second law of thermodynamics).

Fewer than a third of the elements known in the universe occur in living matter, but all of these also occur in nonliving matter. Each element exists in the form of characteristic units called *atoms.* An atom consists of a nucleus, containing *protons* and *neutrons,* around which *electrons* move in orbits. The diameter of the atom is many times greater than the diameter of the nucleus. The number of protons (positively charged particles) is the same in all atoms of a given element, and the number of electrons (negatively charged particles) is the same as the number of protons. (Atoms themselves are electrically neutral.) This number is characteristic of the element and is known as its *atomic number.* The number and arrangement of electrons are responsible for the chemical characteristics of each element.

Elements are represented by *symbols,* these being one or two letters (the first capitalized) derived from the name of the element. But chemical symbols have a more specific use than as abbreviations for the name. Each stands for a single atom of the element. Thus, H means an atom of hydrogen; C, an atom of carbon; O, an atom of oxygen; Na, an atom of sodium (Latin, *natrium*); and Cl, an atom of chlorine.

While all atoms of a given element have the same number of protons and electrons they may vary somewhat in number of neutrons (uncharged particles). And, since the mass of either proton or neutron is far greater than that of an electron, the mass of an atom is determined primarily by the number of particles in its nucleus. It now appears that most elements exist in more than one atomic form, the differences being determined by more or fewer neutrons in the nucleus than in the most prevalent form of the element. These forms of an element that differ in number of neutrons and therefore in atomic mass are called *isotopes.* There are, for example, at least three isotopes of carbon: $^{12}C$ (also written C-12), which is the usual form, $^{13}C$, and $^{14}C$, the numbers indicating mass numbers (which are approximately the atomic weights). The last-named isotope, $^{14}C$, is radioactive. The unstable nuclei of radioactive atoms disintegrate into lighter atoms by the emission of alpha, beta, and gamma rays. Such radioactive elements can be detected easily and are therefore useful in biological research as tags on particular compounds. (See Appendix C.)

The electrons occupy shells and subshells of different energy levels, each containing a certain potential number of electrons. The physical space in which an electron actually occurs is called its *orbital,* and since orbitals are of different shapes their numbers and shapes determine the shape of the atom. (The arrangement of elements in the periodic table, which groups elements that have similar properties, reflects electron structures.) The potential number of electrons is two in the innermost shell and eight in the others. The number and arrangement of the outermost electrons determine the number and nature of bonds possible in the union of two or more elements.

When the outermost shell of a given atom contains the potential number of electrons the atom is chemically stable. When the outermost shell has fewer than four electrons the atom tends to give up electrons, becoming electrically positive; when it has more than four electrons it tends to take up electrons, becoming electrically negative. Elements tend toward a series of stable configurations by electron loss or gain, or, alternatively, by electron sharing.

Electron loss or gain results in charged particles called *ions.* These are of two kinds, those that are positive through electron loss (*cations*), and those that are negative through electron gain (*anions*). The attraction between and resultant bonding of ions of opposite charge (*ionic bonds*) results in the formation of *ionic compounds.* The stable outer electron configuration may also be achieved through the sharing of electrons by two atoms. The shared electrons pass around and are attracted by both nuclei. Compounds formed in this fashion, of atoms connected through shared electrons (*covalent bonds*), are called *cova-*

*lent compounds.* With atoms that are not electrically equivalent the bonds formed are intermediate between completely symmetrical covalent bonds and ionic bonds; these are called *polar bonds.*

Carbon, which has four electrons in its outer shell, gains stability through four covalent bonds, and it can form a wide variety of compounds with many other atoms. Carbon atoms can also combine with each other in chains or ring structures. Certain covalent compounds contain double or triple bonds in which two or three pairs of electrons are shared (e.g., carbon dioxide, which contains double bonds between the single carbon and each oxygen; double and triple bonds between carbon atoms in various compounds).

As a result of these electron differences atoms combine with each other in highly characteristic manners and proportions. When atoms of two or more elements combine with each other they form a *molecule,* which is the smallest unit having all the properties of a compound. We represent such molecules in writing by empirical and structural formulas. An *empirical formula* uses combinations of symbols for the elements, with subscript numbers to indicate, if more than one, the relative numbers of atoms of different elements (e.g., table salt, $NaCl$; water, $H_2O$; ammonia, $NH_3$; methane, $CH_4$). The number that represents the combining capacity of an element is called its *valence,* the valence being, specifically, the number of electrons an atom has to gain or lose to stabilize its outer shell. The valence of ions is represented by a number bearing a plus or minus sign to indicate the charge on the ion (e.g., sodium, $+1$; potassium, $+1$; iron, $+2$ or $+3$; chlorine, $-1$; oxygen, $-2$). The valence of an atom that forms covalent bonds is determined by the number of shared electrons (e.g., hydrogen, 1; oxygen, 2; nitrogen, 3; carbon, 4). In a *structural formula* the atoms involved in the molecule are represented in approximately their structural relations to each other, bonds being indicated by lines. These are necessarily two-dimensional representations, though the actual structure of the molecule is three-dimensional. (For examples of structural formulas see Figs. 2.2 and 2.3.)

## PHYSICAL CHARACTERISTICS

Early observations with the light microscope (see Appendix C) suggested that cells consist of heterogeneous structures variously described as granular, alveolar, fibrillar, or reticular, suspended in a viscous liquid capable of changing from a watery condition to a jellylike semisolid. Recent observations with the electron microscope (Appendix C) have revealed details of these structures and have demonstrated the presence of a network of membranes.

Living matter is a complex system in which some compounds are in true solution and others are in colloidal suspension. A *colloid* is a mixture of substances in which the dispersed particles are larger than in a *true solution* but are too small to settle out. Colloidal particles are in the size range of 0.0001 to 0.1 micron ($\mu$) in diameter (a micron being 1/1,000 millimeter), while dissolved particles in a true solution are smaller than $0.0001\mu$. A colloid can change from a liquid or *sol* condition to a semisolid or *gel* condition, and in living matter as well as in some other colloids this change is reversible.

Details of the specific structures (organelles) found in cells will be given in Chapter III; their functional relations will be discussed in Chapter IV.

## CHEMICAL CHARACTERISTICS

Although carbon is characteristic of living matter, it is not unique to it, nor is any other element. Living matter incorporates elements from its surroundings and in a cyclic manner returns these to the environment where they will eventually be reused. Although most of these elements are incorporated from and lost to the environment in the form of inorganic compounds, these compounds, with the exception of water, constitute only a small part of living matter. The elements used by living matter are converted into larger compounds that are characteristic of life.

The study of the chemistry of compounds associated with living organisms began as *organic chemistry*. But organic chemistry is not restricted today to biological compounds. It has become the chemistry of compounds that contain carbon and hydrogen, many of which do not occur in living organisms. On the other hand, there are, of course, inorganic compounds in the cell. The branch of chemistry that deals with all compounds and reactions associated with living organisms is now called *biochemistry*.

In the following discussion of the chemical characteristics of living matter, the elements involved will be considered first, then the inorganic and the organic compounds. The most important inorganic compounds found in living matter are water, carbon dioxide, and acids, bases, and salts. The main groups of organic compounds present are carbohydrates, fats, proteins, and nucleic acids.

**Elements.** The most abundant elements found in protoplasm are oxygen, carbon, hydrogen, and nitrogen. Other elements present in smaller quantities are calcium, phosphorus, chlorine, sulfur, potassium, sodium, magnesium, and iron. Some other elements not generally present that may occur in small quantities are lithium, boron, fluorine, aluminum, silicon, vanadium, manganese, cobalt, copper, zinc, selenium, bromine, molybdenum, cadmium, iodine, and barium.

The nonmetallic elements oxygen, carbon, hydrogen, and nitrogen are the most abundant in living matter because of their contributions to the formation of organic molecules (the largest called *macromolecules*). Only one of these elements, oxygen, is abundant in nonliving matter (Fig. 2.1) though no one of the others is rare. The metallic elements commonly constitute only a small fraction of the cell volume but are essential to many phases of cell metabolism.

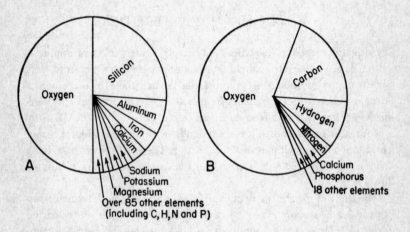

Fig. 2.1. Relative abundance of different chemical elements in the nonliving world (*A*) and in living organisms (*B*). Oxygen is the most abundant element in both; carbon, hydrogen, nitrogen, and phosphorus, also of major importance in living matter, are in relatively low concentrations in nonliving matter. (Reprinted by permission from *General Biology* by Gordon Alexander, published by Thomas Y. Crowell Company.)

**Inorganic Compounds.** There are no inorganic compounds in protoplasm that do not occur in the same form in nonliving matter. The most abundant of these, water ($H_2O$), is a major component of the nonliving as well as of the living world. Living matter is about 75 percent water. Carbon dioxide ($CO_2$) is present as a dissolved gas. Also dissolved in the water of cells are salts, acids, and bases.

WATER. Several of the properties of water are highly significant for living matter: (1) Water dissolves a great many substances, so many, in fact, that it is sometimes called the "universal solvent." (2) Water has a high dielectric constant, which means that electrolytes (see below) in solution in water ionize readily. This property makes water an important medium or agent for chemical reactions. (3) Water has a high heat capacity; hence its temperature changes slowly in relation to changes in its surroundings. A large amount of heat is also required to melt ice or evaporate water. These thermal properties are of great importance in climate control and, through this, control of the distribution and activity of organisms. These properties are also important in stabilizing the internal temperature of organisms. (4) Water expands (decreases in density) just before freezing. Thus ice forms first on the surface of natural bodies of water, insulating the water below. (5) Water has, for a liquid, a high surface tension, its surface in contact with air behaving as a stretched membrane. This property, combined with adhesion between water and soil particles, results in the capillary rise of water in soil. It also makes possible support of small organisms on the surface of a pond. (6) Water has a specific chemical role in many biochemical reactions. It is involved in the synthesis (by dehydration) and the breakdown (by hydrolysis) of large organic molecules (Fig. 2.4). Water is thus essential, not only for living matter itself but for many of the environmental conditions of life as well.

CARBON DIOXIDE. Carbon dioxide, another simple compound, is utilized by green plants in food synthesis and released in all cells as an end product of food breakdown. This compound occurs as a gas in the atmosphere and in solution in natural waters.

SALTS, ACIDS, AND BASES. Many inorganic compounds in living matter are common salts, acids, or bases. Their molecules, when dissolved in water, tend to dissociate into ions, positively charged cations from hydrogen and the metallic elements and negatively charged anions from the nonmetallic elements. Since ions are conductors of an electric current such compounds are called *electrolytes*. (Compounds that do not dissociate and therefore do not conduct an electric current are called *nonelectrolytes*.)

Salts dissociate into various kinds of cations and anions, depending upon the composition of the molecule. Table salt, for example, ionizes into sodium ions, which are positive ($Na+$), and chloride ions, which are negative ($Cl-$). Acids invariably form hydrogen ions ($H+$) and anions that differ with the kind of acid. Hydrochloric acid, for example, ionizes to form $H+$ and $Cl-$ ions. Bases yield various types of cations and, invariably, hydroxyl ions ($OH-$). Thus we say that acids are characterized by hydrogen ions and bases by hydroxyl ions, though we refer to hydrogen ions to indicate concentrations for both (see below).

Water, though not considered an electrolyte, dissociates slightly, forming a neutral solution. If a pure salt is added to pure water the solution is still neutral. But if an acid is added, the concentration of hydrogen ions increases, and, if a base is added, not only does the concentration of hydroxyl ions increase but the hydrogen ion concentration decreases, relatively. Thus the whole acid-base scale can be represented by the hydrogen ion concentration. This is usually done indirectly, however, because the actual concentration would have to be expressed in a rather cumbersome fraction. The value we use, which is expressed by the symbol pH, is the logarithm of the reciprocal of the hydrogen ion concentration. The pH scale runs from 1 to 14. The pH of a neutral solution is 7.0; higher $H+$ concentrations (more acid conditions) are represented by smaller figures, alkalinity by figures larger than 7.0. Since the scale is logarithmic, a pH of 5 is ten times as acid as pH 6; pH 5 is 100 times as acid as pH 7; pH 10 is 1,000 times as basic as pH 7; etc. One cannot average pH figures.

The pH of living matter is near neutrality, and it is relatively constant. This constancy is maintained in part by the presence of certain compounds, *buffers,* that prevent major changes in pH even when small amounts of acid or base are added. Such a compound is derived from a weakly dissociating acid or base. Its buffering effect is due to the fact that addition of $H+$ or $OH-$ ions results in formation of more undissociated molecules that involve these ions; in other words, the concentrations of free $H+$ or $OH-$ ions are maintained at relatively constant levels. Two important buffers in cells are carbonates and phosphates, which are derived from weakly dissociating acids (carbonic and phosphoric acids).

Potassium is the most abundant cellular cation, followed by magnesium and relatively small amounts of calcium and sodium. (In contrast, sodium has the highest concentration in body fluids outside cells, just as it is the most abundant cation in sea water.) The most common anion in cells is phosphate, and carbonate is second; chloride is also present in some cells (e.g., red blood corpuscles). Other anions include sulfate and bicarbonate.

**Organic Compounds.** A great many complex organic compounds occur in living matter. Fortunately for the student, many of these have basic similarities, and a preliminary understanding of living processes requires familiarity with only four categories: carbohydrates, lipids, proteins, and nucleic acids. Furthermore, these organic molecules are composed of a relatively small number of kinds of molecular subunits or "building blocks" (Table 2.1). The large molecules break down into their component subunits on *hydrolysis* (*digestion*), and synthesis occurs through the reverse process of bonding by *dehydration* (*condensation*). See Figure 2.4 for a diagrammatic illustration of these processes in relation to fats.

Table 2.1. Major Types of Biological Compounds and their Subunits. The subunits listed in the left column form the molecules listed in the right by dehydration condensation. The reverse process, hydrolysis (digestion), splits the larger molecules into their component subunits.

| SUBUNITS OR "BUILDING BLOCKS" | LARGE MOLECULES |
|---|---|
| Monosaccharides | Polysaccharides |
| Three fatty acids<br>Glycerol | Fats |
| Two fatty acids<br>One phosphoric acid<br>Glycerol | Phospholipids |
| Amino acids | Proteins |
| Nitrogenous base<br>Pentose sugar<br>Phosphate group | Nucleotides |
| Nucleotides | Nucleic acids |

CARBOHYDRATES. Sugars and the products of their condensation (polysaccharides) are carbohydrates. A sugar consists of a carbon chain (generally arranged in a ring structure) flanked by hydrogen atoms and hydroxyl ($-OH$) groups, the hydrogen and oxygen being present typically in the same proportions as in water (Fig. 2.2). Many carbohydrates are hexose (6-carbon) sugars or yield hexoses when digested. Pentose (5-carbon) sugars occur in nucleic acids and certain polysaccharides and are important in photosynthesis (Chap. IV). Trioses

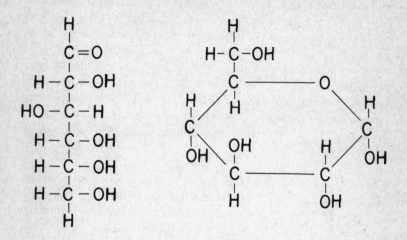

Fig. 2.2. Two different ways of representing the hexose sugar, glucose ($C_6H_{12}O_6$), with structural formulas. The open-chain structure is shown on the left, the ring form or structure on the right. Both, of course, are limited by a two-dimensional representation. Relationships of atoms in a molecule can best be shown with three-dimensional models.

(3-carbon sugars), as well as a variety of other sugars, occur as intermediate products in metabolic activities. The initial carbohydrate manufacture takes place in organisms in the process of photosynthesis. The energy thus stored, which comes from the sun, is then available for biological work through the oxidation of the carbohydrates that are produced (Chap. IV). The intermediate breakdown products of carbohydrate oxidation include molecules that are necessary in the synthesis of other essential compounds. The significance of these will become apparent in Chapter IV.

The carbohydrates are classified as monosaccharides, disaccharides, and polysaccharides.

*Monosaccharides* (simple sugars). Examples: *glucose* ($C_6H_{12}O_6$), one of the key molecules in cellular respiration (Fig. 2.2), and *ribose* ($C_5H_{10}O_5$), a subunit of nucleotides, which form nucleic acids. Some simple sugars with the same empirical formula differ in three-dimensional structure. Such are called *isomers*. Glucose and fructose are isomers; they do not differ in chemical composition (both are $C_6H_{12}O_6$), but they do differ from each other in the arrangement of the atoms in the molecule.

*Disaccharides* (double sugars, consisting of two linked monosaccharides). Examples: *sucrose* ($C_{12}H_{22}O_{11}$), table sugar, which has been commercially removed from plants (that store sucrose in their cells) for hundreds of years; *maltose* ($C_{12}H_{22}O_{11}$), derived in the digestion of starch. Sucrose is formed of one molecule each of glucose and fructose; maltose is formed of two molecules of glucose.

*Polysaccharides* (consisting of numerous linked monosaccharides). Examples: *starch* ($C_6H_{10}O_5$)$_x$, a plant food storage form; *glycogen* (same empirical formula), an animal storage form sometimes called animal starch; *cellulose* (same empirical formula), which forms the rigid cell walls of plant cells and is the chief component of many commercial fibrous products (paper, cotton, linen). Starch has a less branched molecule, and it is somewhat less soluble in water, than glycogen; cellulose, which is insoluble in water, has bonds that are not split by the enzymes acting on starch or glycogen. Mucopolysaccharides are a group of compounds that serve to cement cells together, lubricate joints, and form constituents of blood-group substances.

LIPIDS. The lipids are chemically diverse molecules that are generally insoluble or only slightly soluble in water but are soluble in ethyl alcohol, ether, and some other organic solvents. Three types of lipids commonly occur in protoplasm.

*Fats* (true fats or neutral fats) are glycerol esters of fatty acids; that is, they are compounds of one molecule of glycerol and three molecules of fatty acids (Figs. 2.3 and 2.4). Palmitic acid ($CH_3C_{14}H_{28}COOH$) and oleic acid ($CH_3C_{16}H_{30}COOH$) are commonly occurring fatty acids. Each fatty acid consists of a long hydrocarbon chain that is *saturated* if the carbons contain the maximum possible number of attached hydrogens (as in palmitic acid) and *unsaturated* if any carbon atoms are double bonded, resulting in less hydrogen per carbon (as in oleic acid). Fats occur primarily as food reserve molecules, formed when food material is abundant and degraded when food is in demand. They yield a little more than twice as much energy (calories per gram) as do carbohydrates or proteins.

*Phospholipids* are similar to the true fats, but one of the fatty acids is replaced by a phosphate group, usually with additional water-soluble molecules. The presence of these other groups strongly influences the physicochemical characteristics of the phospholipid, enabling it to bind

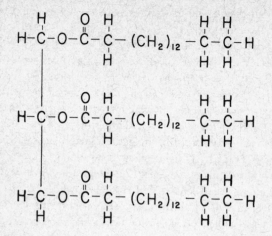

Fig. 2.3. Structural formula of a fat, palmitin. Its structural composition involves condensation of one molecule of glycerol, here shown at the left, and three molecules of fatty acid (in this case palmitic acid), shown extending at right angles to the right. Condensation of the four molecules takes place by dehydration, three molecules of water being lost in the process (Fig. 2.4). The three atoms of carbon and five of hydrogen in the left vertical axis were derived from glycerol, the other atoms from the fatty acids. (The long carbon chains of the fatty acids have been abbreviated within the parentheses.)

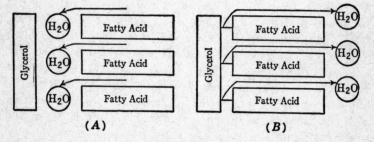

Fig. 2.4. Diagrams illustrating (as they occur in fats) the processes of hydrolysis and dehydration. (A) Hydrolysis (digestion) of a molecule of fat into one molecule of glycerol and three of fatty acids involves the addition of three molecules of water. (B) Dehydration condensation of one molecule of glycerol and three of fatty acids to form a molecule of fat involves removal of three molecules of water.

water-soluble compounds to water-insoluble compounds. This type of molecule is apparently an important component of cell membrane systems. *Lecithin* is a commonly found phospholipid, especially abundant in nerve tissue and in egg yolk.

*Steroids* are structurally different from true fats, having four fused carbon rings with an attached carbon chain of varying length. In small concentrations, various forms of these compounds exert biological regulatory effects on sexual development and function as well as on certain aspects of metabolism in higher organisms. Common examples include *cholesterol,* vitamin D, sex hormones, adreno-cortical hormones, and the bile salts.

PROTEINS. Proteins are large molecules consisting of long, unbranched chains of *amino acids*. Although the twenty or so different types of amino acids ordinarily involved have distinctive structural characteristics, and therefore distinctive chemical properties, they all have the same configuration at one location. This configuration, an amino group ($-NH_2$) and an acidic carboxyl group ($-COOH$) attached to the same carbon atom, is the key to the manner in which adjacent amino acids are linked by *peptide bonds* (Fig. 2.5). These

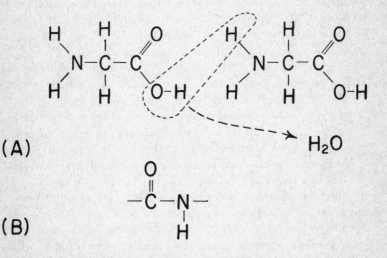

Fig. 2.5. (*A*) Two molecules of glycine, a simple amino acid. These form a dipeptide by dehydration, the process here indicated diagrammatically. Linkage between the two amino acids is by the peptide bond (*B*), formed between the carboxyl group of one molecule—in this case the one shown on the left—and the amino group of the other, that on the right, above.

bonds are formed between the amino group of one amino acid and the carboxyl group of the next, one molecule of water being removed in the formation of each bond. The remaining components of each amino acid extend from the protein as arms consisting of hydrogen (as in *glycine,* the simplest amino acid) or of a variety of straight-chain or ring compounds, which may also contain sulfur or additional oxygen and nitrogen.

Two linked amino acids constitute a *dipeptide*, three a *tripeptide,* and a large number a *polypeptide.* Polypeptide chains form proteins. Because there are twenty different kinds of amino acids that can be linked in any sequence, and because protein molecules can be very large, there is an extremely large number of potential protein types. Each protein is characterized by the kinds, numbers, and sequence of amino acids it contains, its *primary structure.*

The sequence of amino acids forms a spiral, with the hydrogen of the amino groups and the oxygen of the carboxyl groups forming a weak bond (hydrogen bond) that holds the spiral together. This spiral, referred to as the alpha helix, is called the *secondary structure.* Imposed upon this structure is an additional three-dimensional configuration of folding and looping that results from various types of bonding between amino acids. These bonds include hydrogen bonds, ionic bonds (involving electrostatic attractions), and disulfide bonds. (When two sulfur-containing amino acids are in close proximity the sulfur atoms link, forming the disulfide bond, a bridge.) This structure of loops and folds is called the *tertiary structure.* In addition to this folding, some proteins have linkages between two or more adjacent polypeptide chains (e.g., insulin, which consists of two polypeptide chains held together by two disulfide bridges). This type of structure, gained by combining polypeptide chains, is termed by some the *quaternary structure.*

*Native proteins* are those in the presumed natural state; *denatured proteins* have had certain bonds broken and, therefore, the original folding altered. Proteins that have a general rodlike shape are called *fibrous;* those that are folded over at a number of places are called *globular.* *Simple proteins* yield only amino acids on hydrolysis; *conjugated proteins* have nonprotein groups (e.g., metals, polysaccharides, lipids) bonded to the amino acid chains.

Proteins contain both acidic and basic groups, owing to the different types of component amino acids (which can be acidic, basic, or neutral). At any given pH the net charge on a particular protein molecule may be positive, negative, or neutral. Migration of protein molecules in solution in an electric field may, therefore, be either to the positive or the negative electrode. For this reason, proteins are said to be *amphoteric*. (This characteristic behavior aids investigators in separating different protein and amino acid molecules. See Appendix C.) Because they can neutralize acids and bases, proteins are important buffers in body fluids such as blood, which must maintain constant pH.

Almost all cellular differences within an organism and among organisms can be traced to protein differences. Proteins are essential in the formation and maintenance of the structural and functional machinery of cells. They are, after *deamination* (removal of $-NH_2$), a source of energy in the diet, being oxidized much as are carbohydrates. And, owing to the complete interconnection of metabolic pathways, steps in protein degradation provide basic metabolic energy.

Types of proteins include:

(1) Proteins that function in a structural capacity (e.g., *keratins* of fingernails, skin, and hair; *collagens* of connective tissue; lipoproteins of cell membrane systems).

(2) Proteins that are *hormones*, regulators of metabolic processes (e.g., *insulin*).

(3) Proteins that function in the transport of oxygen (e.g., *hemoglobin*).

(4) Proteins that are components of chromosomes (e.g., *histones*).

(5) Proteins that are organic catalysts (*enzymes*), controlling reaction occurrence and rate (e.g., enzymes of digestion, enzymes of cellular metabolism). All enzymes known thus far are proteins. Like other catalysts they activate a reaction but are not used up in the process. Some accomplish their catalytic activity alone, some require a cofactor (certain vitamins are precursors of cofactors), and some must be combined with a metal to function. Enzymes reduce the quantity of energy necessary to activate a reaction; thus, in the presence of an enzyme, the reaction occurs under conditions in which it would not otherwise take place (e.g., at body temperature). The three-dimensional structure, as well as the chemical composition, is essential to enzyme activity, for a denatured enzyme does not function as a catalyst. One essential in the process appears to be the three-dimensional union of the enzyme (plus a cofactor if one is necessary) with the substance acted upon, which is called the *substrate* (Chap. IV).

NUCLEOTIDES AND NUCLEIC ACIDS. Nucleic acids are organic compounds that are even larger than proteins. As with proteins they are composed of repeating subunits. In nucleic acids these subunits are *nucleotides.* Each nucleotide consists of a nitrogen-containing organic base (chemically a purine or pyrimidine), a pentose (5-carbon) sugar (ribose or deoxyribose), and phosphoric acid.

There are two types of nucleic acids, *DNA* (*deoxyribonucleic acid*), which has deoxyribose in the nucleotides, and *RNA* (*ribonucleic acid*), which has ribose as a nucleotide component. Either type of nucleic acid has four possible nucleotides, depending on which nitrogenous base is present. DNA nucleotides contain the purines *guanine* and *adenine* and the pyrimidines *cytosine* and *thymine*. The same possible kinds of nucleotides compose RNA except that thymine is replaced by *uracil*. The sugar of each nucleotide is attached to the adjacent nucleotide through the phosphate, and the purine and pyrimidines are side branches.

The DNA molecule consists of two parallel nucleotide chains in a double helix, with the organic bases extending toward each other. These bases pair, adenine only with thymine, and guanine only with cytosine, with stabilizing bonds (hydrogen bonds) holding the adjacent polynucleotide chains in position (Fig. 2.6). As a result, the strands are *complementary* but not identical. DNA is predominantly found in chromosomes (traces have also been reported in other intracellular structures), and, as such, it performs basic functions in the control of cell expression and in the transmission of hereditary information from one cell to the next cell generation. The theory that the pair of complementary polynucleotides that composes the DNA molecule separates, forming new DNA molecules after the synthesis of new complementary units, gives us a chemical explanation of gene replication (Chap. V).

In contrast to DNA, most RNA is found as a single strand, although in some instances this may be self-reflexed into a helix. There are three types of RNA found within the cell, *ribosomal RNA, transfer RNA,* and *messenger RNA*. In association with DNA, RNA controls protein synthesis. The different forms of RNA are considered in relation to their functions in Chapter V.

Certain modified nucleotides play a vital role in the transfer and storage of chemical energy in cells: *ADP* (*adenosine diphosphate*) and *ATP* (*adenosine triphosphate*). ATP is formed from ADP by the addition of a phosphate group; the oxygen-phosphorus bond, indicated by a wavy line ($-O\sim P$), has a high potential energy level. Conversion of ADP to ATP takes place as energy is stored in the cell; when ATP loses the third phosphate group the stored energy becomes available for energy-requiring cell functions.

OTHER ORGANIC COMPOUNDS. Certain groups of complex organic compounds exist in living matter in small quantities, yet they are essential to the normal functioning of the organism. The *porphyrins* are large ring systems usually attached to a protein and containing a metal. The metal porphyrins include iron-containing *heme,* a component of hemoglobins (important in oxygen transport) and *cytochromes* (involved in cellular energy transfer), as well as the magnesium-containing

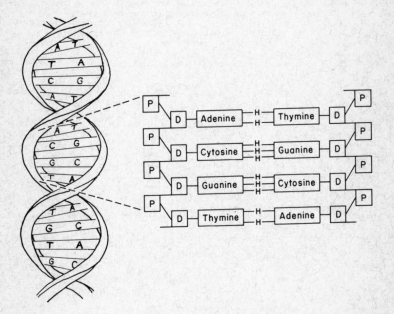

Fig. 2.6. At the left, a diagram of a portion of the double helix composing the DNA molecule. The nitrogen bases of the two rows of nucleotides are indicated by initial letters. Four sets of nucleotides have been projected in more detail, diagrammatically, at the right. Here one sees that the "band" providing continuity down each edge of the helix is formed by successively bonded phosphate-deoxyribose (P-D), while the nitrogen bases are linked across the center by hydrogen bonds. Linkages of nucleotides are possible only between the bases thymine and adenine, and between cytosine and guanine. This relationship suggests the basis for complementary replication of a single strand of DNA (or synthesis of messenger RNA, with uracil rather than thymine as the complement of adenine, this treated in Chap. V). Note that two hydrogen bonds link adenine and thymine, and three link cytosine and guanine.

*chlorophylls* (which absorb light energy during photosynthesis). *Vitamins* are chemically diverse compounds that are required in minute amounts for the activity of certain enzymatic pathways. *Alkaloids* are complex nitrogen-containing compounds (e.g., nicotine, morphine) found in plants, many of which produce physiological effects in man, such as the relief of pain and the production of hallucinations.

# Chapter III

# CELLS: MORPHOLOGY

The cell is the smallest unit of biological activity that displays the special attributes by which we characterize life (Chap. I). The term cell was first used in 1665, by Robert Hooke, who applied it to the regularly occurring empty chambers he saw in thin slices of cork. Today, the term is used for a dynamically changing protoplasmic unit organized within a membrane system. It typically includes a centrally differentiated structure, the nucleus; a surrounding portion, the cytoplasm; and an outer membrane. The cell maintains itself as such by extracting materials and energy from its environment, and it is capable of growing and of duplicating itself.

## THE CELL THEORY

Although Matthias Schleiden, a German botanist publishing in 1838, and Theodor Schwann, a German zoologist publishing in 1839, are generally credited with the first statement that all organisms consist of cells, the cell theory (or cell concept or principle) was actually formulated earlier by several other biologists, including Lamarck (1809) and Dutrochet (1824). The statement by Rudolf Virchow, in 1858, that "where a cell exists there must have been a pre-existing cell" provided the connection between the cell theory and the theory of evolution. During the late 1870's and early 1880's the complex process of cell division was discovered and analyzed by Strasburger, Fleming, and others, and this understanding of cell reproduction led to modern interpretations of development and inheritance. As recognized today, the cell theory states that a living organism consists of one or more cells and their products, and that all cells arise from preexisting cells.

# THE SIZE OF CELLS

Cells vary in size from near the lower limit of visibility with an ordinary light microscope (about 0.5 micron in diameter) to the size of the yolk of an ostrich egg (some 70 mm in diameter). Some cells, such as the nerve cells of certain animals, are long in length, with a relatively small diameter. The significant factor in all cases is the ratio of surface area to volume because this relates to the exchanges between the cell and its surroundings. The smaller the cell, the greater the ratio of surface area to volume.

# THE SHAPE OF CELLS

Cells vary greatly in shape in relation to the special functions they serve. For example, nerve cells and muscle cells are elongated, and skin cells are flattened. Cells freely suspended in liquid surroundings are spherical (because of surface tension). In groups, or against surfaces, inequalities of pressure from different sides result in irregularities in form. Tightly packed cells form polyhedrons.

# CELL STRUCTURE

Although cells are found in a variety of sizes and shapes, their structural complexity appears to be a modification of a generalized design (Fig. 3.1). Cells contain intracellular structures or *organelles* that have specific substructures and functions. The major parts of a typical cell are (1) the surrounding membrane, (2) the *cytoplasm,* and (3) the *nucleus.*

The cells of bacteria and blue-green algae are simply organized, lacking the well-defined nuclear membrane, other complex membrane-bound organelles, and the organized chromosomes typical of other cell types. These simple cells are called *procaryotic cells* in contrast to the more complex *eucaryotic cells* of higher organisms. Certain specialized eucaryotic cells may function without a nucleus, e.g., red blood cells of mammals, but such cells are incapable of self-replication. There are also examples in which a continuous mass of cytoplasm contains many nuclei. Such a mass is called a *syncytium* (in animals) or a *coenocyte* (in plants). The single cells constituting some lower organisms may also characteristically have more than one nucleus.

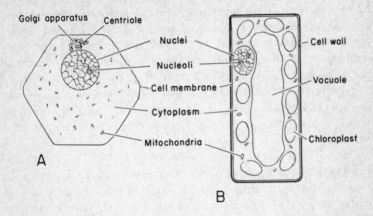

Fig. 3.1. (*A*) Typical animal cell. (*B*) Typical plant cell. Both as seen in "optical section" under the magnification of an ordinary light microscope, approximately × 1,000.

**Cell Membrane.** The outside boundary of all cells is determined by a thin but definite *cell membrane* (*plasma membrane*), which is an active part of the cell. This membrane acts both as a selectively permeable barrier and as an active transport mechanism that can utilize biological energy to move substances through it against normal diffusion gradients. The living cell must carry on a continual exchange of materials with its upon the composition of the molecule. Table salt, for example, ionizes tration gradients and maintaining ionic concentrations both above and below those in the surroundings (for example, nerve and muscle cells continuously pump sodium ions out of the cell while potassium ions are maintained in higher concentration inside than out). Compounds that are produced within a cell but that are normally considered too large in molecular size to pass out (e.g., secretions produced by cells of digestive glands) may be passed out of the cell by the activity of the membrane system.

Before the electron microscope came into use the cell membrane was best known through its functional control of the passage of materials. Observations of its permeability clearly indicated, for example, the presence of lipids in the membrane. Studies of myelin (the membrane of nerve cells) with the electron microscope have revealed some of the details.

This membrane consists of a three-layered structure, 70 to 100 Angstrom units thick (an Angstrom unit is one ten-millionth of a millimeter, or one ten-thousandth of a micron), the composite being referred to as the *unit membrane*. The unit membrane apparently consists of outer layers of protein with a double phospholipid layer between them. Though many biologists think of the unit membrane as typical, cell membranes probably vary considerably in chemical composition. Furthermore, they do not act as completely uniform structures; they have what are presumed to be structural irregularities ("functional pores") through which molecules may be selectively passed. And, finally, the surfaces of the cells of multicellular organisms often have cementlike substances that hold neighboring cells together.

The cell membrane may extend outward to form *microvilli,* or it may fold inward, forming invaginations that may pinch off and become *vacuoles*. The larger vacuoles formed in this manner are *food vacuoles,* as they may include organic material taken in from outside the cell. Other types of vacuoles will be mentioned when we consider the organelles of the cytoplasm.

**Cell Wall.** Plant cells are surrounded by an additional but nonliving layer of material formed by the cell, the *cell wall*. This layer, which is composed predominantly of cellulose, provides support. Although porous, it is not selectively permeable and permits most molecules to pass through. The cell wall consists of an elastic *primary wall* and, in woody portions of the plant, of a variably thickened, ridged, *secondary wall* that contains lignin as well as cellulose.

**Cytoplasm.** The cell region within the cell membrane but outside the nucleus is called the cytoplasm. Under the light microscope this appears as a relatively uniform, semifluid matrix (ground substance) containing solid granules or filaments. In some cells (e.g., *Amoeba*) it consists typically of an inner portion, the granular and fluid *endoplasm,* and an outer region, the clear and somewhat rigid *ectoplasm*.

The advent of the electron microscope has markedly advanced biological investigations of the cytoplasm. Electron microscopy, coupled with modern techniques in cell physiology, has revealed this area of the cell as architecturally complex, with distinctive structural components related to the biological machinery of the cell (Fig. 3.2). These components include the endoplasmic reticulum, ribosomes, Golgi apparatus, mitochondria, lysosomes, plastids, and centrioles.

THE ENDOPLASMIC RETICULUM AND THE RIBOSOMES. The cytoplasm

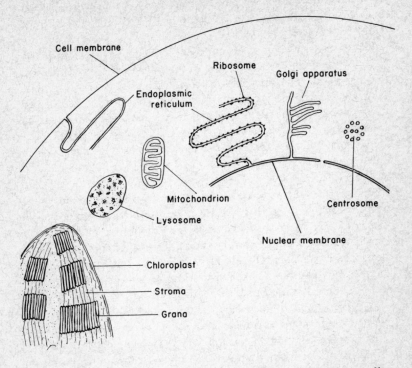

Fig. 3.2. Cellular organelles and parts of organelles as revealed in studies with the electron microscope. These are shown as parts only of cells because to represent all of a single cell properly at this scale of magnification would require a drawing some three feet or more in diameter. Only short arcs of cell and nuclear membranes are shown, and only a portion of a chloroplast. Both types of endoplasmic reticulum, smooth and rough, are represented. The magnifications are approximately 15 to 25 times greater than in Fig. 3.1.

of cells is subdivided by a network of double-membrane structures called the *endoplasmic reticulum*. These membranes form a canal system that is continuous with the nuclear membrane and that is thought to communicate with the exterior through pores in the cell membrane. Thus it may serve as a transport system for materials.

The cytoplasmic region between the reticular units has been called the *hyaloplasm*. Two forms of the reticular membrane occur, *smooth* and *rough* membranes, the latter distinguished from the former by their heavy coating of particles called *ribosomes*. (The endoplasmic reticulum is rather simple in amoebae, whereas it has a highly articulated, rough formation in cells that specialize in protein formation, such as those of the pancreas.)

Ribosomes, dense particles composed of protein (40 percent) and RNA (60 percent), are found free in the hyaloplasm as well as along the endoplasmic reticulum. The ribosome (clusters of which are called *polyribosomes*) is the site where amino acids are assembled and joined during protein synthesis. Investigators have indicated that, in addition to protein synthesis, the endoplasmic reticulum is associated with glycogen storage, cholesterol synthesis, and the conduction of impulses in striated muscle.

THE GOLGI APPARATUS. The Golgi apparatus has been known since 1898, when it was first noted by the Italian cytologist Camillo Golgi. The electron microscope has revealed it as a canal-like system of smooth membranes, probably continuous with the endoplasmic reticulum. The canals border clusters of small vacuoles. The Golgi apparatus is well developed in secretory cells and nerve cells but is small in muscle cells. It apparently participates in the processing of protein secretions; however, its general cellular function is still in dispute.

MITOCHONDRIA. Mitochondria (sing. mitochondrion) are double-membrane structures, the inner membrane having many folds, called *cristae,* that extend into the central cavity or *lumen*. They are spherical to rod shaped and vary from less than a micron to about 5 microns in length. These organelles contain enzymes and are centers of the aerobic phases of cellular respiration and energy-yielding metabolism. (Apparently the capacity to transfer hydrogen to molecular oxygen, as described in Chapter IV, is uniquely located in the mitochondrial membrane system.) Apart from the role of coordinating enzyme activity, mitochondrial membranes display selective permeability, functioning as diffusion barriers. Mitochondria are found aggregated in cells involved in activities requiring high energy expenditure.

LYSOSOMES. Lysosomes are globular cell organelles about the size of mitochondria that contain hydrolytic enzymes. A lysosome is bounded by a membrane that presumably insulates the cytoplasm from the activities of the enclosed enzymes, which catalyze the breakdown of macromolecules. These organelles rupture in injured or senile cells, as well as in cells whose developmental fate is death (e.g., the cells in the tail of a tadpole during its metamorphosis into a frog). In these cases their function is cell breakdown and removal. In intracellular digestion the lysosomes fuse with food vacuoles, forming a vacuole sometimes called a digestive vacuole. In this vacuole food molecules are digested by the hydrolytic enzymes, the end products of digestion diffusing into the cell proper.

PLASTIDS. Plastids are double-membrane organelles characteristic of plant cells and certain unicellular organisms. They include leucoplasts, chloroplasts, and chromoplasts.

*Leucoplasts,* which are colorless, are most common in tissues not exposed to light. They serve as storage units. Those in the cells of roots and tubers often contain starch grains, while those in the cells of seeds accumulate various oils.

*Chloroplasts,* which are the centers of photosynthesis, contain *chlorophyll.* (In blue-green algae, and in bacteria that have chlorophyll, chloroplasts are absent.) The chloroplast double membrane surrounds a system of membranes called *lamellae.* Areas where the lamellae are tightly stacked are called *grana;* where the lamellae are not in close contact the *stroma,* or matrix, often contains starch grains and vacuoles. The chlorophyll is apparently attached to proteins in the grana. The number of chloroplasts in a cell ranges from one (in some algae) to several score (in cells of higher plants).

*Chromoplasts* contain various plant pigments other than chlorophyll (e.g., xanthophylls, carotenes). Although chromoplasts may result from chlorophyll breakdown, unmasking other pigments also present in chloroplasts (e.g., in the process of leaf color change in the fall), they more commonly develop directly, typically producing the red, orange, and yellow colors of flowers and fruits.

CENTRIOLES. Centrioles occur in most animal cells and in the cells of certain plants. They are located in a clear area, the *centrosome,* near the nucleus. Each centriole consists of nine groups of fibers arranged in a cylinder. In paired centrioles, the fibrillar structures are at right angles to each other. The paired centrioles separate during cell division, forming the poles of the mitotic apparatus from which radiate fibers of the spindle and aster (Chap. V). During the last stages of mitosis a new centriole is produced at right angles to each of the two single centrioles. Cilia and flagella, motile hairlike or whiplike extensions of the cell, have at their bases a fibrillar structure similar to that of centrioles. Such structures, which control the motion of the flagella or cilia, are known as *kinetosomes* or *basal granules.* (In the body of a cilium or flagellum the nine peripheral filaments surround two central ones.)

VACUOLES. Vacuoles, regions in the cytoplasm occupied by watery solutions, are of several types. *Food vacuoles* have previously been mentioned in connection with their formation by the cell membrane. Mature plant cells often contain a conspicuous vacuole that, with growth of the cell, gradually displaces the cytoplasm to a thin peripheral layer. This large, fluid-filled structure, which contains dissolved substances, including sugars and salts, has functional as well as structural significance. (It is sometimes called the "sap vacuole," but its contents should not be confused with the sap of trees.) Some unicellular organisms have another type of vacuole, the *contractile vacuole,* which acts to discharge excess liquid from the organism.

**Nucleus.** The nucleus is a specialized body, usually spherical, suspended in the cytoplasm. As observed under a standard light microscope the living cell nucleus is translucent, and it is difficult to differentiate from the cytoplasm. The phase-contrast microscope (Appendix C) makes the nucleus more obvious because of the optical contrast provided by the slightly different refractive indices of the cytoplasm and nucleus. The nucleus becomes a very prominent structure through numerous cell staining techniques using basic dyes (Appendix C).

The substance of the nucleus, the *nucleoplasm,* is continuous with the cytoplasm through pores in the double layered *nuclear membrane.* The outer layer has been observed to be continuous with the endoplasmic reticulum, and both layers have the unit-membrane structure. During cell division the nuclear membrane breaks down, losing its identity (Chap. V).

Suspended in the nucleus is a network of filamentous material, the *chromatin.* Chromatin contains nucleoprotein, the nucleic acid involved being DNA. (The genetic substance of a bacterial cell may consist of a single loop of DNA; in higher organisms DNA is more complexly organized in association with the protein histone.) RNA manufactured in the nucleus is concentrated in definite rounded bodies called *nucleoli* (sing. nucleolus). The nucleoli are chromosome derivatives. They are thought to be involved in the control of protein synthesis. During nuclear division (mitosis) the nucleus loses its identity and the chromatin becomes condensed into characteristic structures, the *chromosomes,* the numbers and kinds of these being uniform in the cells of a given organism. Each chromosome as observed during mitosis consists of two parallel components called *chromatids,* held together by the *centromere,* which also has connections with the mitotic apparatus. Further information on DNA-RNA relationships in cell activity, DNA replication, and nuclear and cell division is presented in Chapter V.

# Chapter IV

# CELLS: PHYSIOLOGY

Cell physiology deals with the functioning of plant cells, animal cells, and microorganisms. The universal similarity in cell morphology implies a basic similarity in physiology; and, indeed, the fundamental cell activities—metabolism, irritability, maintenance of cell environment, growth, reproduction—are similarly performed by all cells. For that reason, an understanding of cell physiology simplifies, and is essential to an understanding of the whole organism. Although there are cellular differences in methods of obtaining energy from foods and in the ability to synthesize macromolecules, physiological differences among multicellular organisms are best reflected in the analysis of the development and operation of organs and organ systems that enable organisms to exist in specific environments. That is the subject matter of Part Two of this book.

## CELL–ENVIRONMENT RELATIONSHIPS

Cells require a relatively restricted range of environmental conditions for their existence and normal functioning. These basic environmental restrictions relate to available energy, the characteristics of compounds required as raw materials, and the regulatory role of factors such as temperature and water. The cell membrane further regulates the conditions within the cell. The irritability of the cell or organism, supplemented by behavioral traits in higher animals, is often essential for proper maintenance within necessary environmental conditions (Chap. XXI). Furthermore, through specific adaptations, cells and organisms have developed abilities to exist under various types of environmental stress.

Living organisms require a continuous supply of energy. This energy, initially from the sun, is transferred to the potential energy of chemical bonds in the process of photosynthesis. All cells are dependent upon the molecules thus produced. Cells may convert this chemical energy to energy (1) for synthesis of other compounds, (2) for movement, (3) for active transport, (4) for heat production, (5) for light production (e.g., in a firefly), (6) for production of electricity (e.g., in an electric fish). In the cell, the high-energy organic compounds are broken down for energy expenditure, and low-energy molecules are released to the environment. Although the elements and molecules released to the environment may be further changed and translocated by biological or geochemical processes they will eventually provide the raw materials for another cycle of biological synthesis. Thus, although energy is constantly in demand the nutrients are repeatedly cycled through living systems.

## TYPES OF NUTRITION

*Autotrophic cells* manufacture their own food, obtaining the energy for the process from the sun (in photosynthesis) or from chemical reactions (in chemosynthesis) and using as raw materials relatively simple inorganic compounds. (These inorganic compounds may have to be in specific forms; for example, nitrogen in the form of nitrate is required by higher plants.) *Heterotrophic cells* cannot manufacture their own food but must get it from other organisms. Many heterotrophic cells or organisms require certain complex molecules, such as amino acids and vitamins, as well as some minerals. These particular dietary requirements are as essential as the food-energy requirements because of their basic contribution to the functional machinery of cell metabolism. Green plants, by photosynthesis, and a few microorganisms, by chemosynthesis, are autotrophic; animals, some plants, and many microorganisms are heterotrophic.

## FUNCTIONS OF THE CELL SURFACE

All nutrient materials must pass through the cell membrane in order to enter the cell. The cell membrane of a living cell is *differentially permeable* ("semipermeable"). This means that certain substances in solution can pass through it while others cannot. The cell wall that surrounds many plant cells, giving them support, is freely permeable.

The various ways the cell membrane regulates the nature and concentration of the cell contents are by dialysis, osmosis, active transport, pinocytosis, and phagocytosis.

**Dialysis.** Molecules tend to move from regions of greater to those of lower concentration when no barrier intervenes. This movement is *diffusion,* and the gradient of concentration from greater to lesser is called the diffusion gradient. The passage of material in solution through a differentially permeable membrane, though it may be in the direction of the diffusion gradient, is a special type of diffusion called *dialysis.* Movement of dissolved substances through the cell membrane takes place by dialysis.

**Osmosis.** Water passes freely through most cell membranes. Since it is the solvent, and since only certain solutes pass through the cell membrane, a difference in concentration of solutes inside and outside the cell may occur. If those solutes that do not pass through the cell membrane are more concentrated inside the cell than outside, water will move into the cell. (The water will be moving from regions where there is more water to regions where there is less, in other words along its diffusion gradient; and there are more water molecules, of course, in those regions where there are fewer molecules of other substances in solution.) This movement of the solvent, water, through a differentially permeable membrane is a special type of diffusion called *osmosis,* and the pressure generated by the differential movement is called osmotic pressure. (Note: The solution with the greater concentration of solutes has the higher osmotic pressure, so movement through the membrane is from the region of lower to that of higher osmotic pressure.)

If material in solution is less concentrated outside than inside a cell, water will move into the cell more rapidly than it moves out. The cell may swell and eventually burst. Under opposite conditions, when solutes are more concentrated outside than inside, water is withdrawn from the cell and it shrinks (Fig. 4.1). Two solutions in osmotic equilibrium have the same osmotic pressure and are said to be *isotonic.* A solution with a higher osmotic pressure than another is *hypertonic* to the latter, while the solution with the lower osmotic pressure is *hypotonic* to the one with which it is compared. Water moves through a cell membrane from hypotonic regions to regions that are hypertonic.

Plant cells in a hypotonic medium, such as cells of plants that live in fresh water, do not burst when water enters by osmosis; the swollen cells press against the cell walls, setting up an internal pressure (*turgor*

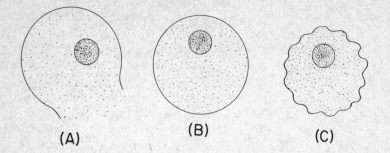

Fig. 4.1. Effects of osmotic pressure gradients on a typical animal cell. (*A*) Exposed to a hypotonic medium the cell swells and bursts, water moving in more rapidly than out. Cell disintegration as a result of this process is sometimes called *laking*. (*B*) Exposed to an isotonic medium the cell's osmotic pressure is in balance with that of its surroundings and the cell does not change in volume; water moves in and out at the same rate. (*C*) Exposed to a hypertonic medium the cell shrinks; water moves out faster than it moves in. This process, which throws the cell membrane into wrinkles, is sometimes called *crenation*.

*pressure*). It is turgor pressure that produces the characteristic rigidity of plant tissue. If a plant cell is placed in a hypertonic solution, however, water is withdrawn from it by osmosis, and the cell membrane shrinks away from the cell wall. This results in the condition called *plasmolysis* (Fig. 4.2).

**Imbibition.** One function of the cell wall other than support, evident particularly in seeds and roots, is the *imbibition* of water. Water may be absorbed and concentrated in the cell wall before it is passed along into the cell.

**Active Transport.** Many materials needed in a cell are required in greater concentration than they occur outside the cell. Also, certain waste products must be discharged from inside the cell at a greater rate than the normal diffusion gradient would provide. Furthermore, certain solutes in the cell fluids are maintained at relatively constant levels even when the concentration of these materials varies in the external medium. To achieve these functions, the membranes must have the ability to transport various materials in specific directions and at variable rates. Such movements do not occur by dialysis because they may be against diffusion gradients. This process of moving materials through the membrane in a direction opposite that of the diffusion gradient is called *active transport*. It requires energy, and the cell membrane plays a role

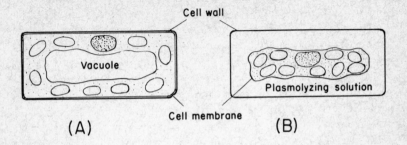

Fig. 4.2. Effects of osmotic pressure gradients on a typical plant cell. (*A*) When exposed to a hypotonic (or isotonic) medium the cell membrane is in contact with the inner surface of the cell wall. This is the normal appearance of the plant cell. Lowered osmotic pressure outside, as when plant cells are placed in pure water, does not change their appearance but produces an increase in turgor. (*B*) When exposed to a hypertonic solution the cell shrinks, pulling the cell membrane away from the cell wall. This effect on the plant cell is called *plasmolysis*. Note that the vacuole has disappeared.

in converting chemical energy into this type of work. (Active transport is also involved in the functioning of the membranes of intracellular organelles.)

**Pinocytosis and Phagocytosis.** Not all material that enters a cell passes directly through the cell membrane. The membrane develops structural modifications that may be involved in such transport. Minute droplets of the surrounding medium may be engulfed by folds of the surface (this process being *pinocytosis*); or solid particles may be incorporated in vacuoles at the surface, these moving into the cytoplasm (this process being *phagocytosis*). Phagocytosis is the method of feeding characteristic of many heterotrophic microorganisms such as amoeboid protozoa (p. 87).


# CELL METABOLISM

The metabolism of a living cell is its total chemical activity, which maintains a dynamic equilibrium while the constituents of the cell are undergoing degradation and replacement. Synthetic, energy-demanding reactions of metabolism are called *anabolism;* and the reverse, breaking-down, energy-releasing reactions are called *catabolism*.

**Enzymes and Coenzymes.** The reactions occurring within the cell require the presence of *enzymes*. Enzymes are biological *catalysts,* compounds that accelerate chemical reactions, enabling the cell metabolic machinery to operate under its relatively low temperature (the temperature of living organisms), low pressure (atmospheric pressure), and relatively restricted pH range. An enzyme catalyzes a reaction by reducing the amount of energy necessary to activate it, but the exact way in which this is done is still unknown.

An enzyme is a protein with a complex, three-dimensional molecular structure (Chap. II). During the reaction in which it is involved it combines with the compound on which it acts, the *substrate,* forming an *enzyme-substrate complex.* The enzyme is released unchanged at the end of the reaction and is able to initiate this type of reaction again.

Reactions succeed each other so rapidly that even a low enzyme concentration has a profound effect on the rate of chemical activity. (One enzyme molecule may, in some cases, activate more than a million substrate molecules per minute!)

The union of the enzyme and substrate takes place at a position on the enzyme referred to as the *active site.* This three-dimensional union has been considered analogous to a lock and key or mortise and tenon type of association. *Enzyme inhibitors,* the so-called metabolic poisons, are thought to interfere with or eliminate normal enzyme-substrate activity by blocking or otherwise influencing the active site.

Some enzymes require an additional compound, such as an organic cofactor, or a metallic ion in order to be active. There are two types of cofactors, those tightly bound to the enzyme (*prosthetic group*) and those loosely bound (*coenzymes*). The cofactor, if not synthesized by the organism that requires it, must be included in the diet. (Most vitamins are compounds from which cofactors are derived.) The inactive protein component of the enzyme is called the *apoenzyme.* The term *holoenzyme* is sometimes used for the combination of apoenzyme and coenzyme.

The action of an enzyme is typically limited to one type of chemical reaction. Its substrate is a particular compound or, in some cases, any one of several closely related compounds. Names of most enzymes end in *-ase* and indicate either the molecule acted upon (e.g., sucrase, which digests sucrose) or the type of action (e.g., decarboxylase, which removes carbon dioxide from a molecule). The enzyme action may be reversible, depending upon relative concentrations of substrate and end products and upon availability of the necessary energy. Each enzyme operates at an optimum pH, and enzymes are destroyed by high temperature (the temperature of boiling water).

In recent years the discovery has been made that enzymes may exist in slightly different forms that vary in enzymatic activity and environmental requirements. These closely related enzymes, which are called *isozymes,* are found in variable quantities in different parts of multicellular organisms and in different developmental stages. They may play significant roles in differentiation.

**Major Metabolic Processes.** Metabolic reactions operate in sequence, the product of one reaction being the substrate of the next, and each serving as an important link in the total chain of reactions. Of the numerous metabolic reactions taking place in the cell, certain sequences can be singled out as particularly significant. Likewise, the compounds or enzyme-substrate complexes that interconnect metabolic sequences have pivotal importance.

The metabolic sequences can be summarized under four categories: photosynthesis, respiration, biosynthesis, and digestion (Fig. 4.3). These

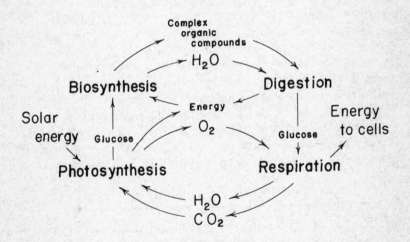

Fig. 4.3. Diagram illustrating the relations among the four major metabolic processes. Glucose manufactured in photosynthesis is condensed in biosynthesis to complex organic compounds; these are digested to glucose; the glucose is then oxidized in the process of respiration. Energy acquired from the sun in glucose manufacture (photosynthesis) is released in the cell in the process of respiration. The glucose manufactured in photosynthesis can, of course, be immediately oxidized, without passing through the steps of biosynthesis and digestion, thereby providing energy for biosynthesis as well as for other cellular activities. Photosynthesis and biosynthesis are endothermic (endergonic) reactions, digestion and respiration are exothermic (exergonic).

will be described below, followed by a discussion of the more important metabolic intermediates. The biochemical sequences involved in photosynthesis and respiration will then be given in more detail in later sections of this chapter.

Photosynthesis and biosynthesis are both anabolic reactions. They are *endothermic* (*endergonic*), which means that more energy goes into the reaction than is released. In other words, potential energy is stored. Digestion and respiration, on the other hand, are catabolic reactions. They are *exothermic* (*exergonic*), releasing more energy than is required to produce the reaction.

Although oxygen is the final oxidizing agent in catabolism, many biological *oxidation-reduction reactions* occur without oxygen, involving oxidation by the removal of hydrogen or loss of electrons. Furthermore, the hydrogen atoms and electrons do not accumulate but are donated to an acceptor. In this manner, the electron or hydrogen donor is the reducing agent (that becomes oxidized in the process) and the electron or hydrogen acceptor is the oxidizing agent (that becomes reduced in the process).

PHOTOSYNTHESIS. Food production takes place in chlorophyll-containing cells by the process of photosynthesis. The energy involved is from light (as implied by the root "photo"), and, in natural photosynthesis, this light is sunlight. The raw materials used are water and carbon dioxide—from the environment. The net effect, chemically, is reduction, leading to the formation of simple carbohydrates and accompanied by the release of free oxygen. The process is complex, involving numerous enzyme sequences to be described later in this chapter, but the total process may be abbreviated in the following equation:

$$6CO_2 + 12H_2O + \text{energy (solar)} \rightarrow C_6H_{12}O_6 + 6H_2O + 6O_2$$

The fact that the released oxygen comes from water rather than carbon dioxide has been verified by tracer experiments using the oxygen isotope $^{18}O$. Light energy produces a splitting of water, releasing hydrogen atoms and oxygen, the "light reactions"; the hydrogen is used to reduce carbon dioxide to glucose, the "dark reactions." Although light initiates other types of cellular reactions, resulting in such responses as vision, phototaxis, and photoperiodism, the energy is not converted into a gain in potential energy as it is in photosynthesis but is used immediately.

RESPIRATION. Although respiration may require accessory activities in some organisms (e.g., breathing), the process is fundamentally an activity of the individual cell. As such, respiration is a complex series of intracellular reactions, releasing the energy stored in food molecules during photosynthesis. The process, chemically, is *oxidation,* and the net effect is the reverse of photosynthesis. The complete process requires oxygen; the original raw materials of photosynthesis, carbon dioxide and water, are given off as end products. The food molecules are oxidized by the removal of hydrogen, which may be joined with the oxygen to form water. The type and amount of respiration is controlled by the presence or absence of oxygen in the environment.

Although only a few cells carry on respiration exclusively in the absence of oxygen (*anaerobic respiration*), the first reaction sequences in glucose oxidation are the same whether in the presence or absence of oxygen. In the presence of oxygen (*aerobic respiration*), these reaction sequences lead to the *citric acid cycle* (Krebs cycle) rather than to the formation of lactic acid or ethyl alcohol (fermentation). The citric acid cycle, receiving molecular units from all major food materials, completes the breakdown, giving off hydrogen and carbon dioxide. *Oxidative phosphorylation* results in the combination of the hydrogen with oxygen, yielding energy stored in ATP and giving off water.

BIOSYNTHESIS. In biosynthesis simple food molecules produced by photosynthesis or obtained in nutrition are combined (condensed) to form larger molecules (Table 2.1). Large carbohydrate molecules (polysaccharides), representing stored food condensed from simple sugars, exist principally as starch (in higher plants) and glycogen (in animals and some plants). Stored fats of many different kinds are condensed from glycerol (derived from a triose sugar) and fatty acids (long-chain compounds derived from many combined acetyl groups). Amino acids and nucleotides are similarly derived from compounds involved in respiratory metabolism. (In various heterotrophic organisms certain groups of these compounds cannot be formed and are therefore required in the diet.) The production of the complex macromolecules of nucleic acids ("informational macromolecules") and proteins is controlled by the hereditary material (see Chap. V).

In these biosynthetic steps, ATP and other nucleotide triphosphates supply the energy for bond formation between "building block molecules" (through what can be considered a *dehydration* process). As an example, the formation of glycogen results from the repeated bonding of activated glucose molecules, each attached to the end of the growing glycogen chain. Although the attachment of each glucose molecule requires a series of reactions involving both ATP and UTP (uridine triphosphate) it can be simplified as follows:

$$nC_6H_{12}O_6 + nATP \rightarrow n \text{ glucose-phosphate} + nADP$$
$$n \text{ glucose-phosphate} \rightarrow (C_6H_{10}O_5)_n + n \text{ phosphate} + nH_2O$$

The cell not only manufactures a variety of large molecules through biosynthesis but assembles them into the structures of intracellular units or organelles. Thus the complex architecture of the cell is produced and maintained.

DIGESTION. In general a cell membrane is impermeable to larger molecules but is permeable to many subunits. Because of this, large molecules must be broken down (digested) for transport and for absorption through cell membranes. During the life of the cell its various molecular components are continuously being built up and broken down. This gain and loss of molecules is dependent upon the rates of synthesis and digestion, and it varies with the cell and the types of compounds involved.

Although digestion involves a sequence of reactions that in net result are *hydrolytic,* and therefore the reverse of synthesis, the specific pathways are not necessarily the reverse of those in biosynthesis. In the processing of food by a heterotroph, specific enzymes that break down various types of organic compounds are secreted into the digestive vacuole, the digestive tract, or the cell environment. The digestive process is a stepwise sequence resulting in the separation of the building block molecules. The process is illustrated by the following equation for the digestion of a polysaccharide such as starch or glycogen:

$$(C_6H_{10}O_5)_n + nH_2O \rightarrow nC_6H_{12}O_6$$

This is accompanied by release of energy, which may be channeled into ATP production if typical intracellular metabolic pathways are involved.

**Pivotal Metabolic Intermediates.** Several types of compounds, essentially unaffected in metabolism, are involved as intermediates because they act as carriers of energy or compounds that interconnect important metabolic sequences. Important metabolic intermediates include ATP, a basic energy storage and energy transfer molecule; NAD and NADP, which are carriers of hydrogen; and coenzyme A, which is a carrier of acetyl groups.

ATP. Of major importance in accepting, storing, and releasing energy in cell metabolism are the modified nucleotides ADP (adenosine diphosphate) and ATP (adenosine triphosphate). In cellular metabolic activity, energy flows from high-energy donors through ATP to energy-requiring reactions in the cell. When energy is in excess, ADP accepts the phosphate ions, storing energy in the form of ATP, which acts then essentially as an energy reservoir. When chemical work (synthesis), transport work (active transport), or mechanical work (muscle contraction) is done, the terminal high-energy phosphate is donated to the proper acceptor, which releases the phosphate ions to the cell after energy expenditure, and ADP is produced (Fig. 4.4). Although other components also act as energy-carrying intermediates, the ADP-ATP system is the pivotal intermediate in cell energy transformations.

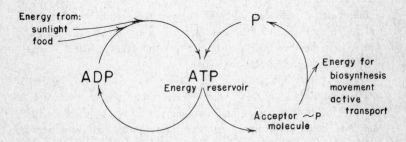

Fig. 4.4. Energy in the cell is stored primarily by the continuous synthesis of ATP (adenosine triphosphate) from ADP (adenosine diphosphate) and phosphate. This energy is subsequently released for cellular functions by conversion of ATP back to ADP.

The production of ATP is called *phosphorylation*. It can occur through the direct utilization of solar energy (photophosphorylation) or through the energy release gained by cellular oxidation of organic compounds (oxidative phosphorylation). Since organic compounds have been initially constructed by solar energy, the ADP-ATP system results essentially in the transfer of energy from the sun to various energy-expending systems of cellular activity.

NAD AND NADP. As mentioned previously, biological oxidation-reduction reactions often involve the removal and addition of hydrogen (or electrons). NAD and NADP are key hydrogen-transfer intermediates that oxidize by the removal of hydrogen. These transfer intermediates are then restored to their original form by the donation of the hydrogen to other compounds. The transfer often involves electrons (the hydrogen being split into the nucleus, $H^+$, and the electron, $e^-$), but since the end result is the transferal of hydrogen the process is called either hydrogen or electron transfer. Both NADP, *nicotinamide adenine dinucleotide phosphate* (previously known as *TPN, triphospho-pyridine nucleotide*), and *NAD, nicotinamide adenine dinucleotide* (previously known as *DPN, diphosphopyridine nucleotide*) are nucleo-tide derivatives that contain the vitamin nicotinamide of the B complex.

In photosynthesis, light energy results in the splitting of water, the hydrogen passing to NADP. The hydrogen is eventually used for the biosynthetic reduction of carbon dioxide to carbohydrate. Thus, along with ATP, $NADPH_2$ links the light and dark reactions of photosyn-thesis. NAD is a hydrogen acceptor-carrier that oxidizes a variety of carbohydrates through hydrogen removal. In the presence of oxygen, the hydrogen is generally passed by $NADH_2$ from organic compounds to the oxidative phosphorylation chain, with the eventual production of water. (In some cases, a flavoprotein, *FAD, flavin adenine dinucleotide,* which is a member of the oxidative phosphorylation chain, directly accepts and transfers the hydrogen.)

COENZYME A. Coenzyme A (usually written *CoA*) is a carrier of acetyl groups ($CH_3CO-$). During cell respiration, major types of foods (glucose, fatty acids, and some amino acids) have their carbon skeletons broken down into 2-carbon or acetyl pieces that combine with CoA, forming *acetyl CoA*. Acetyl CoA is a key compound that provides fuel, in the form of acetyl groups, to the citric acid cycle. In this manner, coenzyme A forms a pivotal link between major foodstuffs and the ter-minal respiratory process of the citric acid cycle.

**Photosynthesis.** The process of photosynthesis is commonly divided into two reaction sequences, the light reactions and the dark reactions, the former requiring energy from light and the latter not de-pendent on light energy. In the light reactions, water is split into hy-drogen and oxygen by solar energy, with the production of ATP. The hydrogen is transferred to NADP, which, as $NADPH_2$, provides the reducing power for the conversion of carbon dioxide to carbohydrates in the dark reactions; the ATP provides the energy (Fig. 4.5).

Chlorophyll is essential for the light reactions. Several types of chlorophyll occur. One of these, chlorophyll *a,* is present in all green plants and is the significant compound in absorption and transfer of light energy. This type of chlorophyll absorbs light in the red and violet portions of the visible spectrum. There are other types of pigments in

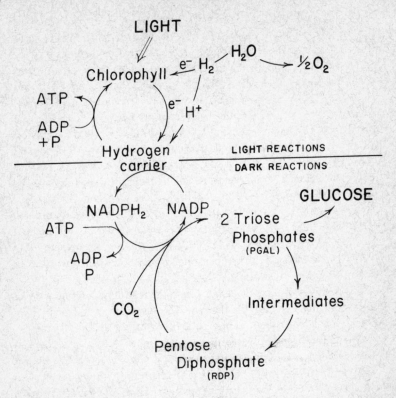

Fig. 4.5. The process of photosynthesis consists of two sets of reactions. The first, the "light reactions," require energy from light; the second, the "dark reactions," do not require energy from light. These are represented in condensed fashion in the above diagram, the light reactions above the horizontal line, the dark reactions below. Chlorophyll absorbs the light energy and decomposes the water molecule during the first steps. The oxygen released at this stage is one of the end products of photosynthesis. Photophosphorylation, formation of ATP directly with energy from light, occurs as one phase of the light reactions. During the dark reactions the hydrogen from water, and carbon dioxide, are combined with pentose diphosphate to form compounds that yield glucose plus intermediate products that eventually reconstitute pentose diphosphate. The pentose diphosphate thus reconstituted can combine with hydrogen and carbon dioxide to form more glucose.

the chloroplasts, and these may be involved in absorbing additional light energy (of other frequencies) and transferring this energy to chlorophyll *a*.

Light Reactions. Light energy absorbed by chlorophyll *a* causes its electrons to move from orbitals near the nucleus to higher energy, outer orbitals. These high-energy electrons are transferred to a series of electron acceptors; as they are cycled back to the original level (the ground state) the energy released is used to produce ATP from ADP and phosphate. This reaction sequence is called *cyclic photophosphorylation*.

Other excited electrons follow a different pathway in which they are passed to NADP, thus reducing it. The electrons lost by chlorophyll *a* are replaced by electrons derived from water; ATP is also formed, and this reaction sequence is called *noncyclic photophosphorylation*. With the loss of electrons from water, the molecule splits, liberating oxygen ($O_2$) and hydrogen ions ($H^+$). The hydrogen ions, along with the electrons from chlorophyll, combine with NADP, forming reduced $NADPH_2$. (The net result is the same as it would be if water, on splitting, passed $H_2$ directly to NADP.) With the production of $NADPH_2$ in noncyclic photophosphorylation, hydrogen can now be carried to $CO_2$ in the dark reactions (Fig. 4.5).

The light reactions of plants other than bacteria and blue-green algae occur in the chloroplast grana. The physical organization of the light reaction molecules is vital to their function.

Dark Reactions. Carbon dioxide combines with a pentose (5-carbon) diphosphate, specifically *ribulose diphosphate* (*RDP*), eventually forming two triose phosphate molecules, *phosphoglyceric acid* (*PGA*). These molecules receive energy (from ATP) and hydrogen (from $NADPH_2$), forming a reduced triose phosphate called *phosphoglyceraldehyde* (*PGAL*). Two of these reduced triose phosphates can be converted into glucose essentially by the reversal of glycolysis (see below). Thus, one molecule of a hexose sugar (glucose) is formed from one molecule of a pentose sugar (ribulose diphosphate), the carbon gain coming from $CO_2$ (Fig. 4.5).

The dark reaction sequence also contains a cyclic component, because a ribulose diphosphate is necessary for each $CO_2$ and thus must be regenerated each time. For every six RDP molecules and six $CO_2$ molecules, twelve PGAL molecules result. Two of these are used to form glucose, and the remaining ten can be cycled through a complex series of reactions to restore the original six RDP molecules. In photosynthetic plants these reactions apparently occur in the chloroplast stroma, but certain phases may occur in cells other than autotrophic ones.

**Chemosynthesis.** Some autotrophic bacterial cells obtain the necessary energy for the synthesis of complex organic molecules in chemical form from oxidation reactions. These cells are therefore called chemosynthetic autotrophs. Energy may come from the oxidation of hydrogen, iron, sulfur, or nitrogen compounds.

**Respiration.** The oxidation of glucose proceeds in a stepwise manner; it involves key intermediates derived from the oxidation of other organic compounds. Thus glucose oxidation forms a central role in cellular respiration. In the absence of oxygen (under anaerobic conditions) glucose is converted to pyruvic acid (*glycolysis*), and this is converted to lactic acid by muscle, other animal tissues, and some microorganisms in a process called *fermentation* (Fig. 4.6).

Although there are other pathways, the same basic series of reactions in the anaerobic breakdown of glucose to pyruvic acid also constitutes the initial series of reactions for respiration in the presence of oxygen (under aerobic conditions). Under such conditions the pyruvic acid is converted to acetyl and carried by acetyl coenzyme A to the *citric acid cycle*. This cycle involves loss of carbon dioxide and hydrogen in a definite sequence of reactions that terminate in reproduction of the same molecule that initially combined with the acetyl form. The hydrogen removed is lowered in energy level and eventually combined with oxygen to form water. Some of the energy released in this process, which is called *oxidative phosphorylation,* is used in the production of ATP from ADP and phosphate.

ANAEROBIC RESPIRATION (GLYCOLYSIS AND FERMENTATION). Although this sequence involves a series of chemical reactions, it can be considered in four interconnected phases, of which the first three constitute what is called glycolysis:

(1) Two ATP molecules supply the "starter energy" through the stepwise manufacture of a hexose diphosphate (the glucose being changed through molecular rearrangement to fructose, specifically fructose diphosphate, which has two high-energy phosphate bonds). This activated 6-carbon unit is then split enzymatically into two triose (3-carbon) units, phosphoglyceraldehyde (PGAL).

(2) These triose phosphates (PGAL) are then oxidized by the removal of hydrogen (NAD acting as the hydrogen acceptor). This process, accompanied by the uptake of inorganic phosphate, forms a triose diphosphate (diphosphoglyceric acid).

(3) The succeeding steps result in the production of pyruvic acid and two ATP molecules for each triose diphosphate, one of the intermediates being PGA (phosphoglyceric acid).

(4) The pyruvic acid is reduced by the addition of hydrogen from $NADH_2$, producing lactic acid. This step is characteristic of anaerobic respiration, because the $NADH_2$ would lead to oxidative phosphoryla-

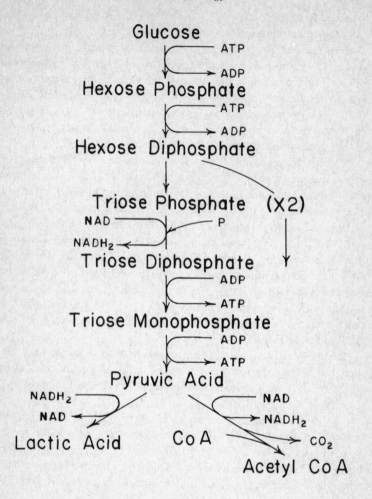

Fig. 4.6. Diagram illustrating the steps in glycolysis—step-by-step conversion of a molecule of glucose into two molecules of pyruvic acid. It begins with phosphorylation, which supplies the starter energy. The hexose molecule is then split into two trioses. (Note: the diagram, to be quantitatively correct, should consist of two, duplicate columns from this point on. This is indicated by "×2" above the vertical arrow.) Pyruvic acid may be converted into lactic acid, under anaerobic conditions, or into acetyl (which combines with coenzyme A), under aerobic conditions. Fig. 4.7 takes up the steps in aerobic respiration where this diagram leaves off, beginning with the acetyl CoA molecule. Oxidation of the NADH₂ is illustrated in Fig. 4.8.

tion under aerobic conditions and the pyruvic acid would be oxidized to acetyl form (Fig. 4.6). This sequence results in a net gain of two ATP molecules per glucose molecule. (The energy yield from anaerobic respiration is about five percent of that from aerobic respiration.)

These reactions occur throughout cells, and, although glucose and lactic acid can penetrate cell membranes, the phosphate intermediates are unable to do so. Under anaerobic conditions lactic acid is commonly produced in cells of higher animals. Lactic acid produced in muscle is carried by the blood to the liver, where it can be converted back to glucose.

In fermentation by yeasts and other microorganisms, glucose is utilized to produce pyruvic acid as in the first three steps above, but the pyruvic acid is then converted to ethyl alcohol and carbon dioxide. Other microorganisms produce acetone and butanol and even utilize compounds other than glucose as an anaerobic fuel source, producing a variety of additional compounds.

AEROBIC RESPIRATION (CITRIC ACID CYCLE AND OXIDATIVE PHOSPHORYLATION). Under aerobic conditions, glucose breakdown follows the same steps as in glycolysis through the production of pyruvic acid. At this point pyruvic acid undergoes oxidation rather than reduction. This can occur because the anaerobic reducer of pyruvic acid, $NADH_2$, has passed hydrogen to the oxidative phosphorylation pathway. The 3-carbon pyruvic acid (formed as the aerobic end product of glycolysis) is oxidized to a 2-carbon acetic acid that is in the activated acetyl coenzyme A form. In this reaction the hydrogens are removed to NAD and the carbon is lost as $CO_2$ (decarboxylation). (Fig. 4.6.)

*The Citric Acid Cycle (Krebs Cycle).* The acetyl coenzyme A donates the 2-carbon acetyl group to oxaloacetic acid (a 4-carbon, dicarboxylic acid), forming citric acid (a 6-carbon, tricarboxylic acid) and releasing coenzyme A. This initiates a chain of reactions that involve carbon and hydrogen loss and that are considered cyclic because they terminate in the production of the original compound, oxaloacetic acid (Fig. 4.7).

The cycle can be subdivided on the basis of carbon removal ($CO_2$ loss), hydrogen removal (to NAD), or a more detailed analysis of the specific chemical intermediates. Considering carbon removal there are three stages, characterized by the six, five, and four carbon compounds:

(1) The 6-carbon tricarboxylic acids, starting with citric acid, are rearranged. (These changes, taking place by the removal and addition of water molecules, include the aconitic acid and isocitric acid mole-

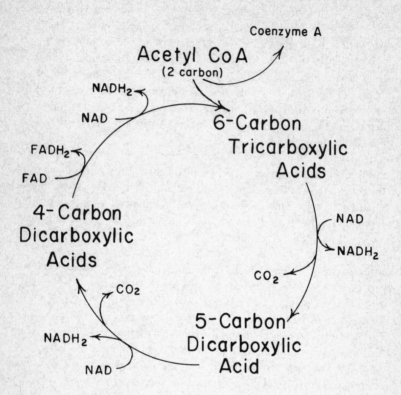

Fig. 4.7. The citric acid cycle (Krebs cycle) begins with the formation of citric acid from a 4-carbon acid and acetyl, the latter passed to the citric acid cycle by coenzyme A. The citric acid is broken down, step by step, to the original 4-carbon acid by dehydrogenation (loss of hydrogen) and decarboxylation (loss of carbon dioxide), releasing energy that converts ADP and phosphorus into ATP. The hydrogen picked up by the reduced NAD is destined to be passed along a chain of acceptors, indicated in Fig. 4.8. Carbon dioxide, released in the formation of acetyl and in the citric acid cycle, is an end product of respiration.

cules.) The end product (oxalosuccinic acid) is broken down, giving off carbon dioxide and hydrogen (carried by $NADH_2$).

(2) The 5-carbon dicarboxylic acid produced (alpha-ketoglutaric acid) is further broken down, giving off carbon dioxide and hydrogen (carried by $NADH_2$).

(3) The 4-carbon, dicarboxylic acid formed (succinic acid) leads to a series of reactions involving other 4-carbon compounds (fumaric acid and malic acid) and resulting in the loss of hydrogen, which reduces both flavoprotein (see below) and NAD. This sequence ends with the production of oxaloacetic acid. The net result has been the oxidation of the acetate, with the removal of eight hydrogens and two molecules of carbon dioxide. The reaction of oxaloacetic acid with another acetyl coenzyme A reinitiates the cycle.

In this manner, the oxidation of a very large number of acetate groups can be brought about by one molecule of oxaloacetic acid because it is regenerated with each "turn" of the cycle. The enzymes of this cycle are located within the soluble matrix of the mitochondria.

*Oxidative Phosphorylation.* The major energy production during respiration is associated with the aerobic fate of reduced NAD, $NADH_2$. This initiates a series of reactions that result in the transport of hydrogen (electrons) to oxygen (Fig. 4.8). The sequence has also been called the *hydrogen* or *electron transport system,* the *cytochrome pathway,* or the *respiratory chain.*

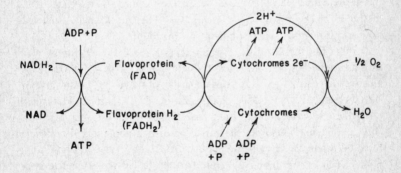

Fig. 4.8. Hydrogen transport. The hydrogen released in various steps in glycolysis and in the citric acid cycle is passed along by a series of acceptors. The last cytochrome in the series reduces oxygen, with the formation of water. Water, of course, is one of the recognized end products of respiration. Most of the energy released in the process of respiration is released in this series of steps.

The reaction sequence can be considered in three interconnected phases associated with the major types of carriers (Fig. 4.8):

(1) The reduced NAD ($NADH_2$) reduces flavoprotein, and in this

process enough energy is released to bring about the formation of ATP. Because the flavoprotein has received the hydrogens, NAD is now oxidized and ready to accept more hydrogens.

(2) The flavoprotein (specifically *flavin adenine dinucleotide,* abbreviated *FAD,* containing as a building block vitamin $B_2$ or riboflavin) can be reduced either by $NADH_2$ or succinate (a 4-carbon, dicarboxylic acid of the citric acid cycle). The reduced flavoprotein, $FADH_2$, is then reoxidized in the process of reducing one of the cytochromes.

(3) The succeeding oxidation-reduction reactions, involving a series of cytochromes, result in the production of two more ATP molecules and end when *cytochrome a* reduces oxygen, forming water.

Thus, in these step-by-step reactions, hydrogen removed from carbohydrate eventually reduces oxygen, resulting in the production of water and ATP. A single reduced NAD molecule (transferring the equivalent of two hydrogens) brings about the formation of three ATP molecules from ADP and phosphate. Each turn of the citric acid cycle brings about the removal of the equivalent of eight hydrogens, which pass down the chain to oxygen, resulting in a total production of twelve molecules of ATP.

The oxidative phosphorylation enzymes are directly associated with the inner membranes of the mitochondria. Their position is apparently fixed and is related to their sequence in functioning.

Aerobic respiration of glucose produces a much higher energy yield than anaerobic respiration. Thus, the complete cellular oxidation of one glucose molecule ($C_6H_{12}O_6$) can produce: (1) two ATP from glycolysis (the same as under anaerobic conditions); (2) six ATP, through oxidative phosphorylation, from the two $NADH_2$ produced in glycolysis; (3) six ATP from the oxidation of two pyruvic acid molecules to acetyl coenzyme A (also linked by $NADH_2$ to oxidative phosphorylation); (4) twenty-four ATP from two "turns" of the citric acid cycle. In summary, 38 ATP are produced by the series of reactions. Aerobic respiration results in an overall efficiency of about 40 percent, with the remaining 60 percent released as heat.

# IRRITABILITY

All cells have the capacity to respond to certain stimuli in their environment. They may be sensitive to mechanical pressure, to differences in light intensity, and to chemical, temperature, and electrical changes.

Responses to these stimuli take two forms, *transmission* (*conduction*) of a change, referred to as an *impulse,* and some type of response by the cell, perhaps involving movement (contraction) of part or all of the cell. Various aspects of irritability will be considered later (in Chaps. IX, XII, and XXI), but here are two illustrations of the dependence of irritability functions on cellular metabolism, one involving conduction, the other contraction.

In a normal nerve cell, which is adapted especially for conduction, the cell membrane is electrically polarized, with an excess of positive charges outside. The electric potential represented by this gradient across the cell membrane is called the *resting potential*. It is due to an unequal distribution of ions on opposite sides of the membrane, specifically the greater concentration of sodium ions outside than inside. This difference is maintained by active transport, the energy for which is supplied by ATP in the nerve cell. When a nerve is stimulated, that point on the membrane is depolarized, which means that the surface of the membrane at that point is electrically negative to all other points along the nerve cell membrane. The depolarization of the membrane is immediately restored, and the membrane again becomes capable of conducting an impulse. The significant point here is that the mechanisms by which the resting potential is maintained and polarization is restored require energy, the energy that goes into active transport, and this energy is derived from cell metabolism.

The other illustration has to do with contraction. In cells that contain contractile fibers, two proteins, actin and myosin, are involved in the process of contraction. For contraction to occur these must be combined in what is called actomyosin. But actomyosin alone will not contract. It must have a source of energy, and the initial energy for this contraction comes from ATP (though in muscles there is also present an additional source of high energy, phosphocreatine). It is worth noting, too, that mitochondria, the cytoplasmic organelles so important in oxidative phosphorylation, are particularly abundant in muscle cells.

In general, the functions of irritability, whether involving one cell only or many cells in complex interrelations, are as dependent upon cellular metabolism as are any other functions of the body. In all cases, the necessary energy comes from the oxidation of foods (which were manufactured by autotrophic organisms, using energy fundamentally from the sun).

# Chapter V

# CELLS: NUCLEAR CONTROL, DIFFERENTIATION, GROWTH, AND REPRODUCTION

Among the most important attributes of living matter are growth, differentiation, and reproduction. These functions, like metabolism and irritability, reside in the individual cells, but these particular attributes are governed by and strikingly dependent upon constituents of the nucleus.

## THE ROLE OF THE NUCLEUS

A single cell, in its normal environment, is in dynamic equilibrium with its surroundings. It retains its characteristic form and function even while in continuous change, taking in materials from the outside and discharging others. Responsible for this continuous metabolic activity is the complex enzyme system. And the enzymes—indeed all proteins—are synthesized in response to chemical instructions in the nucleus of the cell.

Biologists have known for many years that the nucleus of the cell contains the hereditary information that determines the nature of a given organism. They have known, too, that such information is in the chromosomes and that the functional units of heredity, the *genes,* are arranged in linear fashion along the chromosomes. Recently the role of genes as controllers of protein synthesis was established. The site of the stored information was found to be in the DNA (deoxyribonucleic acid) molecules, of which the genes are composed. Though many details must still be worked out, the mechanism by which DNA controls cellular activities, through protein synthesis, is now fairly well established.

**DNA Replication.** As indicated in Chapter II, DNA exists in the form of double strands—two strands of DNA coiled about each other in a double helix, with the pairs of nitrogen bases attached centrally by hydrogen bonds. The two strands are complements of each other, a particular purine base on one strand being capable of bonding only to a particular pyrimidine base on the other strand (Fig. 2.4). Adenine bonds only to thymine, and guanine only to cytosine.

This structure, proposed in 1953 by Watson and Crick, provides the basis for *replication* (*duplication*) of the DNA molecule. The hydrogen bonds between the base pairs break and the double helix unwinds; each of the two strands thus separated serves as a *template* on which a complementary strand is formed. In the presence of DNA polymerase, the new strand is synthesized, nucleotide by nucleotide, of units containing the appropriate complementary purines and pyrimidines. Kornberg and his associates, in 1967, showed this to be the mechanism of DNA replication. Biologically active DNA was produced when natural DNA (serving as template), DNA polymerase, and the necessary raw materials were all available. (At the time some overly enthusiastic newspapers reported that life had been synthesized!)

**Protein Synthesis.** Although this mechanism of DNA replication is involved primarily in cell division, providing continuity from one cell generation to the next, it gives a clue to the mechanism of DNA activity in cell metabolism. DNA functions in the transfer of information from nucleus to cytoplasm by way of ribonucleic acid (RNA) molecules. These molecules of RNA, referred to as *messenger RNA* (mRNA), are synthesized along the DNA strands by the formation of complementary strands in the same way that DNA duplicates itself. Uracil, which occurs in RNA instead of the thymine of DNA, is the complement of adenine. Besides this difference in pyrimidine bases, RNA differs from DNA in containing ribose instead of deoxyribose, and in being primarily a single-stranded structure rather than a double helix.

The sequence of nucleotides in a particular messenger RNA molecule reflects the sequence in a segment of DNA. Since the gene is apparently a segment of DNA, the information in the gene is thus *transcribed* into a messenger RNA molecule, which can leave the nucleus and move to the site of protein synthesis.

The messenger RNA that moves from the nucleus into the cytoplasm becomes associated with a ribosome, in some cases with a group of ribosomes called a polyribosome. Here the messenger RNA serves as a template that determines the linear sequence of amino acids in the particular polypeptide being synthesized. The amino acids are brought to the site of this synthesis, one by one, attached to molecules of *transfer RNA* (tRNA), a form of RNA of lower molecular weight than messenger RNA and sometimes referred to as soluble RNA. (See Fig. 5.1.) Transfer RNA exists in at least as many different forms as there are different kinds of amino acids, each kind serving to transfer one particular amino acid to the forming polypeptide chain. The peptide

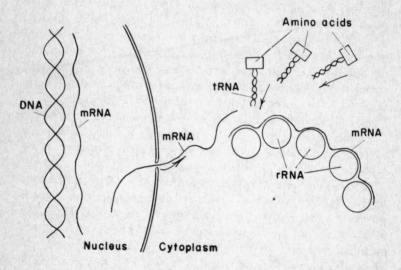

Fig. 5.1. Diagram illustrating the locations and actions of the three types of RNA (mRNA, messenger RNA; rRNA, ribosomal RNA; tRNA, transfer RNA). Messenger RNA is formed by transcription from DNA. It moves out of the nucleus into the cytoplasm, where it becomes associated with the ribosomes (which are about 60 percent ribosomal RNA). Here, transfer RNA molecules bring amino acid molecules, positioning them in sequence corresponding to the coded information in the messenger RNA, and these link together to form a polypeptide chain.

linkages between successive amino acids are brought about by specific enzymes, ATP providing the necessary energy, and the stable, three-dimensional forms are acquired as a product of the amino acid sequence.

Whereas the formation of messenger RNA from DNA is called *transcription,* the synthesis of protein by means of messenger RNA is known as *translation.* (The duplication of DNA is called *replication.*)

In addition to messenger RNA and transfer RNA, there is the form of RNA in ribosomes called *ribosomal RNA* (rRNA). Its exact function is unknown, but it occurs in association with proteins and there is some evidence that it is involved in protein synthesis.

**Genetic Code.** The different sequences of nucleotides in DNA molecules provide the clue to the nature of the information (chemical instructions) carried in these molecules. It is now established that the *code* transferred from DNA to RNA is made up of triplets, "three-letter words." A particular amino acid is represented by a particular triplet— a particular sequence of three nucleotides. This sequence of three bases (for the differences depend upon the nitrogenous bases) in the messenger RNA is called a *codon*. Its complement in the transfer RNA, which translates it as a particular amino acid, is the *anticodon*. The messenger RNA that carries the code for a particular polypeptide chain apparently has a length of several hundred nucleotides.

With four different kinds of nucleotides (four possible "letters") there are potentially 64 different three-letter "words" ($4^3$). But there are only 20 different amino acids. Among the 64 possible triplets, we know that from two to as many as six different combinations may translate as the same amino acid, and that three triplets do not specify any amino acid but apparently function as "punctuation" in the set of instructions, indicating the end of a message. Most triplets code for a particular amino acid but in a few cases the codons are ambiguous, translating in these cases as different amino acids under different environmental conditions. The triplets transcribed are not overlapping, so the transcription must begin at a definite place on the DNA molecule.

The codons are usually indicated in the literature of molecular genetics as three-letter words, the letters being the initial letters of the four nitrogenous bases that occur in RNA. Thus UUU, the first codon discovered and the one that "broke the code" (Nirenberg in 1961), is a triplet in which uracil is the base in three successive nucleotides. This is translated as the amino acid phenylalanine. (Under certain conditions it is translated as leucine.) UUC, as well as UUU, is translated phenylalanine; UUA or UUG means leucine; UCC, UCU, UCA, or UCG means serine; UGC or UGU means cysteine; and UAC or UAU means tryosine. Any one of the four combinations beginning GG (GGG, GGA, GGC, GGU) is translated glycine. The correlation between codons and amino acids is referred to as the *genetic code* (Chap. XVII), and this code is assumed to be universal among organisms. The order of the bases in the codons has been worked out by Nirenberg and others.

A series of nucleotide triplets is apparently what biologists have previously called a gene. Although it has been stated that one gene determines one enzyme, some biologists prefer the term *cistron* to refer to the nucleotide sequence that specifies the amino acid sequence of a single polypeptide chain. (This cistron may determine only part of an enzyme if the enzyme's protein complement has two or more polypeptide chains.) If what we have called a gene determines one enzyme, a gene may consist of one or several cistrons, depending upon the number of polypeptide chains in the enzyme. Some biologists have proposed, too, that a section of the DNA molecule controls the RNA transcription of a series of cistrons ("ordinary" genes or *structural genes*) associated with a specific metabolic pathway. This controlling region is called an *operator gene,* and the total unit of controlled and controlling DNA is the *operon.* Ordinarily, however, the operon does not continuously direct the synthesis of an enzyme. In some cases an enzyme is synthesized only when its substrate is present. And enzyme production ordinarily stops when the enzyme has reached a certain concentration. Analysis of this observation has led to postulation of a *regulator gene* that, in combination with a metabolite produced from the enzyme, represses the action of the operator gene. Thus we have regulator genes affecting operator genes, which, in turn, control the structural genes. The 1965 Nobel prize winners, Jacob and Monod, demonstrated this relationship.

In summary, the nucleus controls cell functions by the following sequence: Information contained in DNA molecules, coded as triplets of nucleotides, is transferred to messenger RNA molecules. These RNA molecules pass from the nucleus into the cytoplasm, where they become associated with ribosomes. There they provide the template for protein synthesis, the amino acids being brought in, one by one, by transfer RNA molecules. Which protein is produced depends on the sequence of codons on messenger RNA. In this manner the sequence of nucleotides in the DNA molecules in the nucleus determines protein synthesis and thus controls the enzyme systems of the cell.

## CELLULAR DIFFERENTIATION

It should be apparent that if DNA molecules are kept identical through self-replication, and if the operation of the mechanism of enzyme synthesis continues unchanged, the descendents of a particular cell will all be alike. Actually, the descendents of a particular unicellular organism are, with rare exceptions, alike; but pronounced differences occur among the cells of a multicellular organism (Chap. VII). These differences are the result of *differentiation*.

In this process the cells of a multicellular organism become specialized for a variety of functions, some even losing their ability to divide and most of them losing their ability to produce new organisms. The cellular differences involve both structural and functional characteristics, including the production of different isozymes (Chap. II) and enzymes, reflecting the differential activity of genes. Experiments indicate that differentiation is not simply the result of a specific sequential pattern of gene activity. Many cells that show differentiation can undergo dedifferentiation, again becoming generalized.

Differentiation also involves interactions between nucleus and cytoplasm and interactions between the cell and its environment. As stressed previously, cell functions are dependent upon environmental conditions, including the supply of necessary raw materials. Metabolic sequences such as protein synthesis can be modified by environmental factors such as differences in temperature and light. As previously mentioned, certain types of protein syntheses appear to have a feedback control in which the production of a quantity of a particular protein inhibits further production, and in certain microorganisms (and conceivably other cell forms) an even more pronounced environmental control results in the production of *induced enzymes,* enzymes produced only in the presence of the proper substrate. In multicellular organisms hormones, too, have a controlling effect on cell metabolic pathways, or even on death and degradation.

Through cytoplasmic influence on the nuclei, the cells that result from the division of certain types of fertilized eggs or zygotes show different developmental capabilities. (See Chaps. X and XIII for specific information on the development of higher plants and animals.) As the multicellular organism develops, the positions of individual cells become important; material produced by certain cells influences other cells (see p. 123 for the effect of auxin on cell elongation and p. 203 for a discussion of embryonic *induction*). Thus, differentiated cells, through chemical influence, can change the fate of other cells. This sequence of events in cellular differentiation suggests such a complex pattern in the development of multicellular organisms that it is amazing that development occurs with so few abnormalities.

## CELL GROWTH AND REPRODUCTION

Growth of an organism involves both increase in the size of cells and increase in the number of cells. Growth in a highly differentiated cell may not be followed by division, but in unicellular organisms and certain cells of multicellular organisms the cell divides after it has reached a particular size. The two resultant daughter cells then begin a new cycle of growth.

The cell, through its enzyme system, is in dynamic balance with its environment, but this does not mean that its anabolic and catabolic activities are in balance with each other. If reactions that result in the synthesis of new materials (anabolism) exceed those resulting in the breakdown of materials (catabolism), the cell increases in volume. An increase in volume may be due, in some cases, to an accumulation of storage products or to the intake of a large amount of water, but true growth involves an increase in the actively functioning macromolecules that make up the cell.

A growing cell eventually reaches a size approximately double that at which it began to grow. During the growth period the number of molecules has increased. The chromatin has doubled in volume, and with this doubling there has been an exact replication of the DNA molecules present in the young cell. The mature cell now divides, forming two cells. During this process, each chromosome splits lengthwise, and each *daughter cell* receives the same number of chromosomes (and the same kinds of DNA molecules) that characterized the *mother cell* at the beginning of the growth period.

Cell division is, of course, cell reproduction. It provides continuity from one cell generation to the next while, at the same time, increasing the number of units. The daughter cells thus formed behave as "young" cells, and they provide a better surface-volume relationship. (Many cell functions depend upon exchanges through the surface, and the increase in volume of a cell appears to have an upper limit above which the surface area is too small and division occurs.)

An individual unicellular organism may have a relatively short life, but its ancestors extend as a continuous line into the remote past, as its descendents, produced by cell division (usually called *fission* in this case), will extend into the future. A similar continuity holds for the cells in a multicellular organism. All its cells are lineal descendents of the first cell that composed it. They have merely remained attached, following cell division, instead of separating as individual organisms. The cells so produced differentiate into a variety of forms, however; only certain ones, the reproductive or *germ cells,* have the ability to produce a new organism, and it is these cells only that have the same type of continuity from the past into the future that is so apparent in unicellular organisms.

When a single cell divides the reproductive process obviously has involved only one parent, the mother cell. Reproduction from a single parent is called *asexual reproduction.* If a new organism arises as a result of the fusion of two cells (e.g., a sperm cell and an egg cell) it has had two parents. This type of reproduction is *sexual reproduction.*

Ordinary cell division is asexual reproduction. This is the typical method of reproduction of unicellular organisms as well as of individual cells of multicellular organisms. Occasionally, unicellular organisms fuse, exchange characteristic DNA molecules (chromatin), and separate. Each newly separated individual has had two parents, in the genetic sense. Each of the separated individuals usually divides immediately; all their progeny have had a double origin and have characteristics of both parental cells. Such a process is obviously a special form of sexual reproduction.

Sexual reproduction in multicellular organisms does not involve an exchange of material between two cells but a complete fusion of the two. The result is the combining of all material in one cell, including, of course, the DNA present in the chromosomes. If ordinary body cells were to fuse, the chromosome number, and therefore the number of DNA molecules, would be doubled each time the process occurred. Such doubling does not take place in the fusion of mature reproductive cells, however, because of a special process that has occurred prior to their formation. This process is a type of cell division that produces cells with half the chromosome complement (and therefore half the DNA). Fusion of two such cells then restores the characteristic chromosome number. The process of division that results in halving the chromosome number is called *meiosis.*

# MITOSIS

Cell division is typically accompanied by a form of nuclear division in which each of the two daughter cells receives a set of chromosomes of the same number and kind as the mother cell had. The chromosome number, which is the same in all individuals of a given kind of unicellular organism and in all body (somatic) cells of a particular multicellular organism, is kept uniform by a process of doubling prior to division. If the chromosome number is 46, as in man, when a cell divides each of the daughter cells receives 46 chromosomes because each of the original chromosomes has duplicated itself.

Ordinarily, division of the nucleus is accompanied by division of the cell. However, biologists sometimes distinguish between nuclear division and division of the cell body, the term *karyokinesis* applying to the former and the term *cytokinesis* to the latter. Together they constitute what is commonly meant by *mitosis,* but many biologists prefer to limit that term to karyokinesis.

**Phases (Steps) of Mitosis.** Mitosis should be thought of as a continuous process, beginning with a single growing cell and ending when the cell has become two independent ones. For convenience of discussion, however, the process is commonly divided into four phases: *prophase, metaphase, anaphase,* and *telophase.* The period between successive divisions is called *interphase.* These phases are essentially the same for plant and animal cells though there are some variations among different organisms. The process as outlined below occurs in most organisms (Fig. 5.2).

PROPHASE. This includes all changes in the cell from the beginning of division, as visible with the light microscope, to the establishment of the chromosomes on the *equatorial plane* of the *spindle* (Fig. 5.2B).

The following major occurrences during prophase are not sequential but rather simultaneous or overlapping in time: (1) The chromatin in the nucleus condenses to form distinct chromosomes, each consisting of two parallel *chromatids* connected by a *centromere* (kinetochore). (2) The two centrioles separate and move to opposite poles of the nucleus. (Centrioles are absent from the cells of higher plants.) At the same time, fibers begin to appear in the cytoplasm, radiating from the centrioles if these are present. (3) The nuclear membrane disappears. (4) The spindle is formed. This consists of two types of fibers, *continuous fibers,* extending from pole to pole, and *chromosomal fibers,* extending from the poles to the centromeres of the chromosomes. (5) The nucleoli disappear. (6) The chromosomes move to the equatorial plane of the spindle.

METAPHASE. Metaphase is the stage in which the chromosomes are on the equatorial plane and during which the separation of the daughter

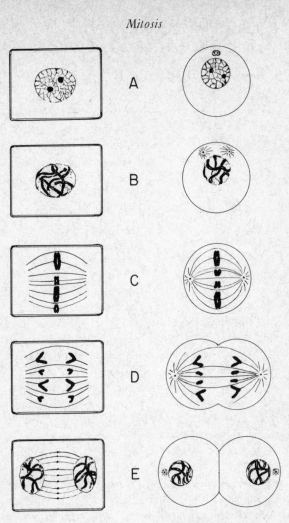

Fig. 5.2. A generalized diagram of mitosis in plant cells (left column) and animal cells (right column). (*A*) Interphase. (*B*) Prophase—condensation of the chromosomes, disintegration of the nuclear membrane, and, in animal cells, migration of the centrioles. (*C*) Metaphase, with the spindle made up of continuous fibers and chromosomal fibers, the latter attached to the centromeres. The chromatids, in the equatorial plane, are beginning to separate. (*D*) Anaphase—migration of the daughter chromosomes to the poles of the spindle. Their deformation is apparently due to the tension of the chromosomal fibers. (*E*) Telophase—organization of the daughter nuclei and separation of the daughter cells. Separation of the daughter cells in plants begins with formation of the cell plate, by coalescence of the granules on the spindle; in animal cells, it takes place by constriction.

chromatids of each chromosome begins. The first event in this separation is the splitting of the centromere (Fig. 5.2*C*).

ANAPHASE. The chromatids of each chromosome separate and migrate to the poles of the spindle—the positions of the new nuclei. They move as if pulled by shortening of the chromosomal fibers. The chromatids are now called daughter chromosomes (Fig. 5.2*D*).

TELOPHASE. In this stage, the chromosomes are transformed into the chromatin network of the interphase nucleus, and the nuclear membranes and nucleoli appear. The centriole, if present, may divide now (or in the next succeeding prophase). If cytokinesis occurs, the daughter cells separate at this stage. In animal cells the spindle fibers disappear and a constriction separates the cells. In plant cells a series of swellings develops on the spindle fibers across the equator; these coalesce to form the *cell plate,* on each side of which a new cell wall is formed (Fig. 5.2*E*).

**Mechanisms of Mitosis.** Several theories exist, none entirely satisfactory, to explain the movements of the chromosomes and the division of the cell on a physicochemical basis. Fibers of some kind are attached to the chromosomes, and these seem to pull the chromosomes toward the poles. Other physical factors that may be involved are diffusion streams and changes in protoplasmic viscosity.

**Division of Cytoplasmic Constituents.** This account of mitosis has dealt primarily with division and separation of the chromosomes, but the major organelles in the cytoplasm also have physical continuity from mother to daughter cells. Division of the centriole has been mentioned in connection with the mitotic apparatus. Mitochondria and plastids separate, so that each daughter cell receives approximately half of those in the mother cell; these structures are individually capable of dividing between successive cell divisions. Cilia and flagella multiply, in association with cell division, by division of basal granules. It is presumed that ribosomes, lysosomes, and the Golgi apparatus also have some physical continuity, although evidence in these cases is incomplete.

# THE CELL LIFE CYCLE

Formerly we knew little of what was going on in a cell when it was not dividing. The cell life cycle was thought of as consisting of mitosis and a period between successive mitoses called the "resting period," or interphase. But the cell is never resting, of course; it is carrying on metabolism continuously. And during part of the period in which we see no visible evidence of division the cell is actually preparing, by synthesis of new DNA, for the next division.

Some biologists have recently suggested that we recognize, as a matter of convenience, a series of four time intervals that occur in sequence in the cell life cycle. These intervals are designated by letters (and subscript numerals) as follows:

$G_1$: *Gap one,* the first period of cell growth, following division and preceding the beginning of DNA replication. (RNA is actively synthesized during this period.) This is the portion of the cell cycle in which a cell remains permanently fixed if it is in differentiated, nondividing tissue.

S: The *synthetic* period of DNA replication. What initiates it is not yet known, but this is the real beginning of the process of mitosis, and once a cell is in this period it normally proceeds to division.

$G_2$: *Gap two,* a period of uncertain length, sometimes essentially nonexistent, in which RNA synthesis continues but DNA replication has been completed.

D: The process of *division,* mitosis. Its first visible evidence is condensation of the chromosomes, their thickening and the establishment of distinct forms.

## AMITOSIS

In some cases nuclear division takes place without formation of a mitotic apparatus. Such division is called amitosis ("not mitosis"). Being negatively defined, the term can and does include more than one type of process. Division of a nucleus by simply pinching in two may occur in cells that do not subsequently divide, or it may even occur in certain specialized or degenerate cells that do divide. Procaryotic cells, like those that comprise bacteria, have a type of "nuclear" division that is similar. The nuclear granule ("chromosome") elongates into a dumbbell-shaped body, then pinches in two, preceding bacterial fission.

## MEIOSIS

The process of meiosis occurs in multicellular animals during the formation of mature germ cells (*sperm* or *egg cells*). In multicellular plants it typically occurs in the formation of special reproductive cells called *spores*. (The relations between spores and the cells involved in sexual reproduction are described in Chapter X.) In both higher plants and animals, meiosis, by which the chromosome number characteristic of the species is halved, occurs at some stage between successive generations, thus maintaining—after sexual union of the cells—the constant chromosome number for the species.

The chromosome number characteristic of a given organism in which sexual reproduction occurs is made up of a double set of chromosomes, each set having come from one of its parents. The chromosomes in each set differ among themselves, varying in size and other morphological features, but when the two sets are combined, as in ordinary body cells, there are two of each morphological type. The two chromosomes of the same type constitute a pair called *homologous chromosomes;* while alike in appearance they do not necessarily carry identical genetic information (Chap. XVII). This double set of chromosomes is referred to as the *diploid number* (2n). (The diploid number in man is 46; in the fruit fly, *Drosophila,* studied so much in genetics, it is 8; in the onion, it is 16.)

Meiosis serves not only to reduce the diploid number to half but to separate homologous chromosomes. Each germ cell (gamete) produced by meiosis contains one of each kind of chromosome, the reduced (*haploid*) number, n, consisting of one chromosome from each of the original homologous pairs. (In man, the haploid number is 23. The chromosomes in the haploid complement are all different, whereas in the diploid complement of 46 there are only 23 different kinds.)

The process of bringing about this reduction, involving separation of homologous chromosomes, requires two successive divisions (Fig. 5.3). It begins with a cell that is essentially prepared for mitosis. The chromatin has doubled, so the chromosomes are ready to divide lengthwise. Instead, in the first steps, the chromosomes unite on the equator of the spindle in homologous pairs. The two homologous chromosomes unite lengthwise in a process called *synapsis*. Since each pair then consists of four chromatids it is called a *tetrad*. The two divisions that characterize meiosis are necessary in order to separate the four chromatids of each tetrad.

The *first meiotic division* results in separating homologous chromosomes. Otherwise the steps are fundamentally as in mitosis. One chromosome of each original pair passes to one pole of the spindle while the other moves to the opposite pole. When this division is

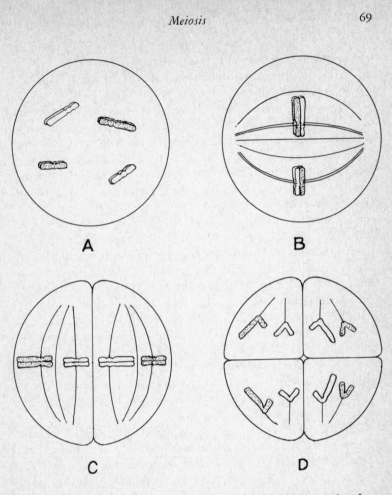

Fig. 5.3. A generalized diagram of meiosis. (*A*) The cell shown has four chromosomes in the diploid complement, a pair of long ones and a pair of somewhat shorter ones. (The stippled chromosomes have come from one parent, the unstippled ones from the other parent.) Each chromosome, prior to synapsis, consists of two chromatids. (*B*) Synapsis. Homologous chromosomes unite, each synaptic pair thus consisting of four chromatids (a tetrad of chromatids). In the division succeeding synapsis each of the two daughter cells receives one chromosome (two chromatids) from each pair of chromosomes. When each of these daughter cells divides (*C*) the chromatids are separated. Thus, each of the four cells formed in the two divisions (*D*) has one chromatid from each tetrad in (*B*). Not only is the chromosome number reduced from the diploid to the haploid condition but in the process each kind of chromosome is retained in each of the four cells.

completed, each daughter cell has only one chromosome (consisting of two chromatids) from each pair of homologous chromosomes (Fig. 5.3C). This first division is therefore sometimes called the "reductional" division in contrast to the second, which is called "equational."

In the diploid complement, each homologous pair of chromosomes consists of one chromosome from the male parent and one, its homologue, from the female parent. In the first meiotic division the individual chromosomes from each parent separate. However, all the chromosomes from one of the original parents do not necessarily move to the same pole. The homologue of male parental origin in one pair may go toward a given pole while the homologue of female parental origin in the next adjacent synaptic pair may go in either the same or the opposite direction. There are thus various possible combinations of maternal and paternal chromosomes in the daughter cells following the first meiotic division. In other words, the chromosomes have been assorted independently of each other. This is a very important principle in Mendelian genetics (Chap. XVII).

In the *second meiotic division* the chromatids of each chromosome separate; so, essentially, we have a mitosislike division that starts with the haploid instead of the diploid number. The net result, after the two meiotic divisions, is four cells from the original mother cell, each with one chromosome (chromatid) from each homologous pair at the beginning. In other words, the four chromatids of each tetrad have been separated in four cells (Fig. 5.3D).

Although meiosis during the production of gametes may be accompanied or followed by cellular differentiation—in the production of sperm and egg cells—the genetic composition remains the same. When two mature sex cells (gametes) fuse, each with the haploid chromosome complement, the diploid number is restored. This process of fusion is called *fertilization,* and the fusion cell is a fertilized egg or zygote. The zygote divides by mitosis in becoming a diploid multicellular organism, but sometime prior to the next generation germ cells in this new individual undergo meiosis, and chromosome reduction occurs.

# Chapter VI

# UNICELLULAR AND SUBCELLULAR FORMS OF LIFE

There are many different types of unicellular organisms, most of them having those properties we used in Chapter I to characterize life and having the organization of a typical cell. There are, in addition, very small entities, the *viruses* (0.005 to 0.275 micron in diameter), that display some of the characteristics of life but are not typical cells; these are here included by our expression "subcellular forms of life."

Unicellular organisms vary greatly in size and complexity. The smallest, only 0.1 micron in diameter, overlap the size range of viruses. These, and viruses, are simple as well as small, because such size sets a limit on the number of molecules present. The simplest unicellular organisms are of procaryotic cells. Bacteria, which are procaryotic, have a usual size range of about one to ten microns in diameter. Though lacking true nuclei, mitochondria, and other typical double-membrane organelles, they have the functional capabilities of eucaryotic cells. Organisms consisting of single eucaryotic cells are extremely diverse in both morphology and physiology. Many have highly specialized organelles in addition to those considered typical of single cells, and some are large enough to be seen by the unaided eye.

The distinction between unicellular and multicellular forms is not always clear. Certain very primitive organisms exist as chains or groups of cells. If the cells are essentially independent of each other we treat the organisms as unicellular. If the cells form an aggregate in which there is specialization of function among them we consider the organism multicellular. But there is a gradient of organization between single-celled and many-celled organisms. The terms unicellular and multicellular are therefore somewhat ambiguous, and in our treatment, particularly of algae and fungi, the distinction may appear arbitrary and merely for convenience.

In this chapter we shall examine typical viruses as well as diverse representatives of unicellular organisms. In Part Two, which deals with multicellular organisms, we shall consider various algae and fungi, the higher plants, and the sponges and higher animals.

# INTRODUCTION TO BIOLOGICAL CLASSIFICATION

Biological classification serves two valuable functions. It groups similar organisms together, and it indicates evolutionary relationships. There are advantages in grouping similar organisms together even when correct relationships cannot be agreed upon. And disagreements can be found among major classification schemes, particularly in their treatment of unicellular and simple multicellular organisms.

**Major Groups.** The initial problem of classification is the division of organisms into the largest groups, called *kingdoms*. Commonly organisms have been placed in either the *plant kingdom* (*Kingdom Plantae*) or the *animal kingdom* (*Kingdom Animalia*)—on the basis of obvious distinctions between multicellular plants and animals. With increased knowledge of microorganisms, however, many biologists have placed the unicellular forms (and, in some schemes, closely related multicellular forms as well) in a separate kingdom (*Protista*). The composition of this group varies with different authorities. A further division, established on the basis of differences between procaryotic and eucaryotic cells, removes the bacteria and blue-green algae from the Protista and places them in a fourth group, *Kingdom Monera*. And some biologists recognize the Kingdom Monera but place all other organisms in either the plant or animal kingdom, recognizing only three kingdoms.

In this outline we shall attempt only to point out major differences in the classification of these groups without adopting one particular system. Rather than consider these classification schemes contradictory we shall use them to illustrate the variety of ways in which living organisms can be compared.

The extremely large numbers of different types of organisms prompt us, in courses in biology, to take advantage of similarities by studying only a few organisms as representatives of groups. Generalized summaries, with concentration only on a few representative types, provide the pattern. We shall here consider the differences in classification as offering an interesting sidelight on such study; these differences need not detract from the analysis of coherent groups of similar organisms.

**Differences among Major Groups.** The major differences separating forms with procaryotic cells from those with eucaryotic cells are as follows:

| PROCARYOTIC | EUCARYOTIC |
|---|---|
| (a) Bacteria and blue-green algae | (a) Other unicellular, and multicellular organisms |
| (b) No nuclear membrane or mitotic apparatus | (b) Nuclear membrane and mitotic apparatus typically present |
| (c) Flagellum, if present, a simple, single strand | (c) Flagella (or cilia), if present, of 2 axial and 9 peripheral strands |
| (d) No true plastids in photosynthetic forms | (d) True plastids present in photosynthetic forms |
| (e) No mitochondria, endoplasmic reticulum, or Golgi apparatus | (e) Mitochondria, etc., present |
| (f) Single-celled or colonial | (f) Single-celled, colonial, or multicellular |
| (g) Cells ordinarily minute, diameter 1–10 microns | (g) Cells usually larger than 10 microns in diameter |

It should be added that although procaryotic cells lack the distinct structures of nucleus, mitochondria, and plastids, there are regions in the cell that carry on the functions associated with these organelles in eucaryotic cells.

The distinctions used to separate the Protista from other kingdoms are variable; some systems place all unicellular forms exclusively in this group, while other systems separate out the Monera. Other versions of the Protista include, in addition to unicellular forms, those multicellular organisms that show such a gradual transition in cellular differentiation that they are clearly related to specific unicellular groups. Because of this ambiguity, particularly in the composition of the Kingdom Protista, we shall emphasize the next smaller category in classification, the phylum. In considering types representative of various groups of organisms we shall, therefore, treat them under the names of the appropriate phyla. As a result of this method, a few of the phyla included in this chapter will appear again when we consider multicellular representatives of the same group.

The following major differences between plants and animals separate many organisms into the plant or animal kingdom:

| PLANTS | ANIMALS |
|---|---|
| (a) Manufacture own food | (a) Require complex foods |
| (b) Have cellulose cell walls | (b) Do not have cell walls |
| (c) Grow throughout life | (c) Grow to a definite size |
| (d) Have no powers of locomotion | (d) Have powers of locomotion |
| (e) Are less irritable | (e) Are more irritable |
| (f) Food stored as starch | (f) Food stored as glycogen |

Many unicellular organisms do not fall into one category or the other because they have some "plant" and some "animal" traits. Such organisms (e.g., *Euglena*) may be considered plants by botanists and animals by zoologists. Furthermore, unicellular organisms typically "grow throughout life"—but that is "to a definite size," at which time they divide. Some multicellular organisms obviously related to plants (e.g., mushrooms) require complex food. The above criteria, as a group, are therefore valid only when we compare complex, photosynthetic, multicellular plants with multicellular animals. There the terms are useful and have a definite place in our vocabulary.

Both plant and animal kingdoms have been divided into subkingdoms. We need here consider only such division of the plant kingdom, however, for when we subdivide the animal kingdom all unicellular animals are automatically separated from multicellular ones. When we subdivide the plant kingdom into its two traditional groups, the Thallophyta and Embryophyta, the unicellular and multicellular forms are still not separated from each other. All the Embryophyta are multicellular (and will be considered in detail in Chapter VIII), but the Thallophyta are an artificial assemblage that includes all primitive plants, some unicellular and some multicellular. The Thallophyta are the algae and fungi. (These terms were formerly used as the technical names of major subdivisions of the Thallophyta. This is no longer the case, but the terms are still used as common names for groups of phyla with similar features.) Algae differ from fungi in having some kind of pigment involved in photosynthesis; fungi, with rare exceptions, obtain their food from outside sources.

The term fungi originally included the bacteria and slime molds but is now generally restricted to one phylum, the Eumycophyta or

Eumycota, the "true" fungi. The true fungi, with some unicellular representatives, are diverse enough to be further subdivided by some authorities. A group of distinct phyla comprise the algae, and single-celled forms occur in all these phyla except the brown algae. (The unicellular red algae will not be considered in this outline.) Some unicellular algae have flagella and carry on active locomotion; some, but not all, have rigid, carbohydrate cell walls. Colonial forms, with a loose association of cells, filamentous forms in which division occurs in a single plane, and more complex forms all develop from single cells that may exist as free-living unicells for various lengths of time.

**Categories of Classification.** Thus far we have considered only major divisions used in classification. But the unit of classification is the *species*. All organisms that are sufficiently alike to have had the same kind of parents belong to the same species. (Although this is approximately correct, more satisfactory definitions of species will be suggested in Chapter XIX.) A particular kind of plant or animal is, in other words, a species. (Note: the word species is spelled with a final *s* in both singular and plural number.) Species of greatest similarity are combined into *genera* (sing., *genus*) ; genera are combined into *families,* families into *orders,* orders into *classes,* classes into *phyla* (sometimes called *divisions* in the plant kingdom), and phyla into *kingdoms.* These terms, for progressively more comprehensive groups of organisms, are the categories of classification. In our consideration we of necessity deal only with types as representatives of the highest (largest) categories, stressing common characteristics used to place different species in the same groups. As previously stated, our emphasis will be at the level of the phylum.

# VIRUSES

Of great biological significance are certain entities whose level of organization includes only some and not all of the characteristics of life, thus placing them on the borderline between living organisms and the nonliving world. These entities, called *viruses,* were first detected through their influence on cell function, often resulting in characteristic diseases of plants and animals. They were initially called filtrable viruses, because most of them, unlike other potentially disease-causing organisms (e.g., bacteria), pass through porcelain filters and are too small to be seen with an ordinary microscope. Viruses are now known to cause the common cold, measles, smallpox, poliomyelitis, rabies, yellow fever, and other animal diseases as well as many plant diseases. The *bacteriophages,* a special group of viruses, are parasitic on bacteria.

In 1935, the virus causing the plant disease tobacco mosaic was crystallized by W. M. Stanley, each crystal representing an aggregate of many particles. The purification through crystallization has aided chemical analysis, and individual viral structure has now been further clarified by use of the electron microscope.

Outside of a cell the virus is inactive. It consists of both protein and nucleic acid components (some viruses also are reported to contain traces of carbohydrate and lipids), the protein forming a sheath around the nucleic acid, which is either DNA or RNA. The viral particles assume various shapes, circular, cylindrical, or brick-shaped; in others (e.g., the extensively studied $T_4$ bacteriophage) the protein sheath has an extension, or tail, with fibers extending from the base.

This extracellular state, when the virus merely has the potential of infection, is ended when the virus attaches to the cell surface (in the case of $T_4$, by the tail) and the viral nucleic acid is extruded into the cell. The different types of viruses are highly specific, attacking the cells only of a specific host. On infection the viral protein is generally believed to be left on the cell exterior (protein functions in various viruses are believed to include protection, support, specific attachment to cell surfaces, and aid in penetration of the cell). Within the cell, the viral nucleic acid behaves as a group of genes that override the cell regulatory mechanism. Using the metabolic machinery of the host cell, these viral genes direct the formation of new viral particles with the same protein-nucleic acid composition as the initial virus. In bacterial cells this eventually results in the destruction of the cell, *lysis,* and the release of numerous virus units. In higher organisms virus particles may travel from cell to cell through cytoplasmic connectives or otherwise be eliminated from the cell without cell degradation. The transmission from infected to uninfected organism may be through the air, as with the common cold virus; through water or on food, as with many insect viruses; or from host to host by insects, as with yellow fever, which is transmitted by the mosquito. (Although the method of transmission was determined by Col. Walter Reed in 1900–1901, the fact that yellow fever is induced by a virus was not known at that time.) An interesting characteristic of some bacterial infections is that the virus at times transports some of the host genetic material to other cells; this is called *transduction.*

A virus does not always cause immediate disease symptoms; it can assume an inactive condition upon cell penetration, generally referred to as the *proviral state*. In this state the viral genetic material replicates, along with the genetic material of the host cells, as if it were added on. The viral nucleic acid may even contribute properties that are advantageous to the host cell, such as immunity from further viral infection. In such a state the virus can be transmitted from cell to cell through the process of normal cell division. During certain types of environmental stress, including the influence of various chemicals and ionizing radiation, the provirus is induced to turn to the reproductive or replicating state, resulting in further viral production and eventual discharge from the cell.

## RICKETTSIAE

Certain extremely small organisms that approach the size of large viruses have also been detected. These organisms, called rickettsiae, are parasitic and disease producing. They are bacterialike in structure but are like viruses in that they reproduce only in living cells. Rickettsiae include the organisms responsible for Rocky Mountain spotted fever (transmitted by a tick) and epidemic typhus (transmitted by the body louse). These have a size range of 0.3 to 0.5 micron.

## UNICELLULAR ORGANISMS

Included in this section are one-celled organisms with either procaryotic or eucaryotic cells. Those with the former type (bacteria and blue-green algae) are often combined in the Kingdom Monera. The others are included either in the Kingdom Protista or distributed between the plant and animal kingdoms, depending upon the authority followed.

**Bacteria (Phylum Schizophyta).** The study of these small and abundant organisms constitutes the economically important field of *bacteriology*.

MORPHOLOGY. The small procaryotic bacterial cells assume rod (*bacillus*), spherical (*coccus*), or helical (*spirillum*) forms (Fig. 6.1*D*) and are usually one to ten microns in length. They may have one or more flagella, but each flagellum has only a single fibril for support.

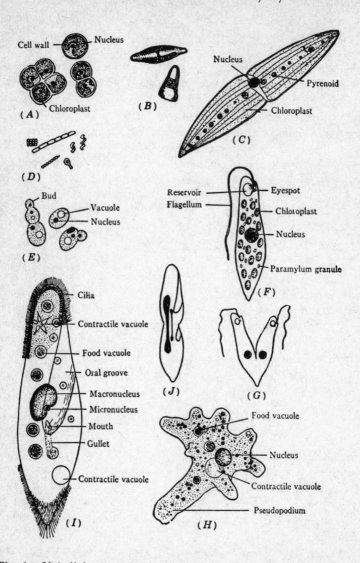

Fig. 6.1. Unicellular organisms. (*A*) *Pleurococcus* (*Protococcus*), Phylum Chlorophyta. (*B*) Diatoms, Phylum Chrysophyta. (*C*) A desmid (*Closterium*), Phylum Chlorophyta. (*D*) Bacteria, Phylum Schizophyta. (*E*) Yeast, Phylum Eumycophyta. (*F*) *Euglena,* Phylum Euglenophyta. (*G*) *Euglena* reproducing by fission, which is longitudinal. (*H*) *Amoeba,* Phylum Protozoa, Class Sarcodina. (*I*) *Paramecium,* Phylum Protozoa, Class Ciliata. (*J*) Paramecium, reproducing by fission, which is transverse.

Although typical double membrane structures are absent, the cells contain an irregular central mass of DNA often called the *nuclear body*. The enzyme systems normally associated with mitochondria and plastids are, if present, associated with inward extensions of the cell membrane. Bacteria have a cell wall composed of both carbohydrates and amino acids (the differences in Gram stain effects, used to separate bacterial types, are due to differences in cell wall composition). The cell wall is frequently surrounded by a slime layer that forms a protective capsule around the bacterium.

PHYSIOLOGY. Although bacterial form is simple, these organisms are biochemically quite diverse. In general, a simple bacterium has most of the metabolic pathways present in other cells, but there are certain critical differences among bacteria. While many of them, called *aerobic bacteria*, require oxygen in their metabolism, others, called *anaerobic bacteria*, develop in complete absence of oxygen. Among the anaerobes are some that function with or without oxygen (*facultative anaerobes*), while others (*obligate anaerobes*) cannot live in the presence of oxygen.

The diverse nutritional capabilities of bacteria are utilized in the classification of different types. Most bacteria are *heterotrophs*. The largest group of heterotrophic bacteria are *saprophytic* (*saprotrophic*), utilizing dead organic material as an energy source. These are important in the decay and recycling of organic material in nature, but certain types are responsible for food spoilage and food poisoning, commonly caused by release of toxin (e.g., in botulism). Other heterotrophic bacteria are *parasitic*, obtaining food from living organisms, and often producing diseases as a result of enzymatic destruction of host tissue or the production of toxins. Cholera, diphtheria, plague, scarlet fever, tuberculosis, and whooping cough are bacterial diseases.

A few species of bacteria are *autotrophs*. A small group restricted to sulfur springs and black muds are *photoautotrophic*. Although these utilize solar energy, the process differs from ordinary photosynthesis in the form of chlorophyll pigments, in the use of hydrogen sulfide or organic compounds as hydrogen donors, and in the fact that molecular oxygen is not a by-product. Another interesting group of bacteria are *chemoautotrophic*, obtaining energy from the oxidation of inorganic hydrogen, nitrogen, sulfur, and iron compounds. Included in this group are the nitrifying bacteria, of significant importance in the nitrogen cycle (Chap. XX).

Other bacteria have special metabolic activities that enable them to obtain nitrogen from the atmosphere (through nitrogen fixation). These forms, of considerable economic value, are either free-living in the soils or are found in association with the roots of leguminous plants.

REPRODUCTION. Reproduction is usually asexual, by fission. Individual bacteria dividing in this manner may remain attached together in chains or masses. Under favorable conditions cell division occurs at the maximal rate of about once every 20 minutes, but, of course, competition for space and nutrients soon limits this rate of division. A form of sexual reproduction, which involves exchange of nuclear material between cells, also may occur in some species.

Although the mutation rate per generation is about the same as that observed in other organisms, the short generation time results in a rapid rate of natural selection and thus the development of new adaptations. Bacteria, because of the short generation time, are therefore very useful in studies of biological inheritance.

Many bacterial forms are able to produce *spores,* which are resistant to brief exposure to boiling water and to desiccation, further increasing the chance of such species for survival.

**Blue-green Algae (Phylum Cyanophyta).** As mentioned previously, the cell form of blue-green algae is procaryotic. The cell wall is similar to that of bacteria. A gelatinous sheath surrounds the cell wall, and flagella are absent. Many forms are colonial, the individual, generally unspecialized cells, produced by asexual reproduction, being held together by the gelatinous sheath (Fig. 6.2). Individual cells are microscopic in size but colonies are readily visible to the naked eye.

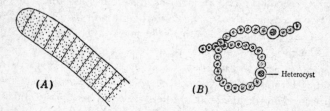

Fig. 6.2. Blue-green algae (Phylum Cyanophyta). (*A*) Tip of a filament of *Oscillatoria.* (*B*) Part of a filament of *Nostoc,* with two heterocysts, large dead cells where breaks in the filament may occur. Both of these forms occur in freshwater habitats.

Blue-green algae contain chlorophyll *a* as well as a blue pigment, *phycocyanin,* and other pigments that may result in red, yellow, blue, or purple color as well as green. (Food is stored as a glycogenlike carbohydrate.) The colors of hot springs are usually due to blue-green algae, some of which live and metabolize at temperatures as high as 70°C. Certain species utilize elemental nitrogen from the atmosphere, contributing usable nitrogen compounds to the soils of shallow water areas such as rice paddies. Others form conspicuous algal "blooms" that may give a disagreeable odor to our drinking water and, occasionally, cause the death of fish or even human illness. Certain blue-green algae are involved in the deposit of marl in lakes. Blue-green algae are widely distributed; they are found in damp soil and in both fresh water and sea water.

**Unicellular Fungi (Phylum Eumycophyta or Eumycota).** Some of the more primitive aquatic fungi produce motile unicellular spores (not comparable to the "spores" of bacteria) that are capable of free-living existence for various lengths of time. The members of a more advanced group, the *yeasts,* economically well known and important unicellular fungi, will be discussed here in some detail.

MORPHOLOGY. Although yeasts typically consist of ovoid or spherical eucaryotic cells (Fig. 6.1E), some form filamentous multicellular hyphae like those of more typical fungi (Chap. VIII). The yeast cell is enclosed in a cell wall composed of polysaccharides (similar to cellulose) associated with proteins.

PHYSIOLOGY. Yeasts are either saprophytic or parasitic heterotrophs. They have relatively simple nutritional requirements and commonly obtain the necessary metabolic energy through the breakdown of sugars. They are *facultative anaerobes;* in other words they do not normally live in the absence of free oxygen but are capable of doing so. In the absence of oxygen, yeasts are able to obtain energy for metabolism by the fermentation of sugar (see Chap. IV). The end products of this fermentation are typically ethyl alcohol and carbon dioxide. The former, in the presence of oxygen, may be oxidized to yield additional energy.

Historically the study of yeast metabolism has made a significant contribution to studies of cell physiology in general. Louis Pasteur, in 1875, demonstrated that fermentation is a cellular phenomenon, and the Buchner brothers gave impetus to the field of enzymatic research 20 to 25 years later by demonstrating the cell-free activity of extracted yeast enzymes, collectively called *zymase.*

The brewer is interested in the alcohol obtained by fermentation of sugars in malt wort, whereas the baker is interested in carbon dioxide production, by which the dough is raised. Wine-making yeasts are obtained from the microbes composing the bloom on the grape skin (in "vintage years" the bloom is dominated by desirable strains) or are inoculated into the purified unfermented grape juice (must). Yeasts are also used to produce enzymes, vitamins, and other useful substances, including various forms of food supplements.

REPRODUCTION. Yeasts reproduce asexually by *budding*. A small swelling (bud) forms on the yeast cell. As it enlarges, the nucleus divides, half going into the bud. A new cell wall then forms between the two cells. Budding may result in a chain, single or branched, of yeast cells. Some yeast species also reproduce sexually, reduction divisions forming haploid cells that combine to form the normal diploid cell.

**Slime Molds (Phylum Myxomycophyta or Myxomycota).** This group is interesting because of the characteristic stages displayed in each life cycle, some of these stages being plantlike and others animal-like. The life history includes a unicellular, flagellated form that is actually a haploid gamete. Under certain conditions pairs of these fuse, forming diploid zygotes. Through subsequent growth, and through nuclear division without division of the cytoplasm, a multinucleate mass called the *plasmodium* is produced. This moves by protoplasmic streaming, and it can feed by phagocytosis. (All stages are heterotrophic.) Under proper environmental conditions the plasmodium ceases to move and develops into a *sporangium*, or fruiting body. The often colorful sporangium, of various shapes, is generally found at the tip of a short stalk. In it, the multinucleate protoplasm produces, by meiosis, uninucleate haploid *spores* with rigid protective walls. (The sporangia and spores are similar, in their significance in the life history, to the corresponding organs and cells in typical plants.) The spores can be dispersed great distances, but under proper conditions they will germinate, giving rise, by mitosis, to the uninucleate cells that are actually flagellated gametes.

**Green Algae (Phylum Chlorophyta).** The green algae have unicellular, colonial, filamentous, and other types of organization. They generally contain chlorophylls *a* and *b,* have starch as a food reserve in chloroplasts, and have cellulose cell walls—all characteristics that link the green algae with the higher plants. This is a large group, with representatives found in a variety of habitats, including moist stones, tree trunks, and snowbanks, as well as in fresh water and, less commonly, marine environments. The unicellular and colonial forms may be flagellated. The following examples illustrate the principles of morphology, physiology, and reproduction as they occur in unicellular green algae. For information on multicellular forms of Chlorophyta see Chapter VIII.

PLEUROCOCCUS (PROTOCOCCUS). The genus name of a microscopic plant that forms green films on the moist bark of trees and the sides of flower pots (Fig. 6.1*A*).

*Morphology. Pleurococcus* is spherical, about 10 microns in diameter, and has a small, central nucleus. One large chloroplast occupies most of the cell. The cell wall is of cellulose.

*Physiology.* Water, oxygen, carbon dioxide, and inorganic salts are absorbed through the cell wall and cell membrane. Nutrition is autotrophic, by photosynthesis. *Pleurococcus* synthesizes its protoplasm from simple inorganic compounds and the sugar it manufactures.

*Reproduction.* The method of reproduction is asexual, by fission. The parent cell divides into two individuals of approximately equal size. Individual plants may remain attached together in groups, but each cell is independent of the others.

CHLAMYDOMONAS. A genus, with numerous species, of motile unicellular algae that occur in standing fresh water or damp soil.

*Morphology. Chlamydomonas* is ovoid, about 20 microns long, with a rigid cell wall (Fig. 6.3). There are two flagella, at the anterior end, and two or more contractile vacuoles near the bases of the flagella. The single chloroplast is cup-shaped and occupies most of the cell; it contains a pyrenoid. A red stigma or eyespot is near the anterior end, at the edge of the chloroplast. There is a central nucleus.

*Physiology.* Nutrition is autotrophic, as in *Pleurococcus.* The pyrenoid, which is protein in nature, is involved in starch formation. The stigma is light sensitive.

*Reproduction. Chlamydomonas* reproduces asexually by fission within the cell wall of the mother cell, after disintegration of the flagella. The daughter cells or *zoospores,* the number produced varying with different species (four shown in Fig. 6.3), acquire flagella and escape from the old cell wall. A similar process results in production of zoospores that behave as gametes, fusing in pairs to form zygotes. In most species these fusing zoospores are alike in appearance (*isogametes*) but belong to different strains (+ or −), as in Figure 6.3, but in at least one species they are unlike (*heterogametes*). Only the zygote is diploid. It develops thick walls and is quite resistant to unfavorable conditions. The first divisions of the zygote are meiotic, resulting in four haploid cells, which become typical zoospores (Fig. 6.3).

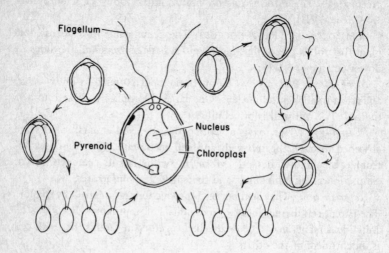

Fig. 6.3. *Chlamydomonas.* Details are shown in the large figure. (The two contractile vacuoles near the bases of the flagella and the dark area in the edge of the chloroplast, the stigma, are not labelled.) Asexual reproduction diagrammatically represented in the cycle at the left; sexual reproduction, at the right. All individuals except the zygote are haploid. The zygote is formed by the fusion of two individuals that behave as gametes; in the case illustrated, these are alike in appearance (isogametes) but from different strains. Meiosis occurs in the zygote, producing four haploid cells that become typical zoospores.

DESMIDS. Desmids are unicellular green algae, which, though of angular or otherwise unusual form, are symmetrical (Fig. 6.1C). They are nonmotile. Individual cells in some species occur attached together in filaments or chains. Desmids are common in freshwater ponds.

*Morphology.* Two symmetrical halves, the *hemicells,* with a median constriction between them, characterize a desmid. A large chloroplast is typically found in each hemicell, and the chloroplasts contain pyrenoids. The nucleus is single, centrally located. There is a cellulose cell wall.

*Physiology.* Nutrition is autotrophic, with typical photosynthesis.

*Reproduction.* This may occur asexually or sexually. In asexual reproduction the nucleus divides, each daughter nucleus entering a hemicell; the hemicells then separate, and each one regenerates another half to complete a new individual. In sexual reproduction two individuals, which look alike, fuse to form a zygote from which a new individual comes.

**Diatoms (Phylum Chrysophyta).** These are unicellular or colonial algae of marine and fresh waters. They do not have flagella, but many species are motile, probably by means of cytoplasmic streaming through slits in the cell wall. Diatom cell walls contain silica; the walls form double "shells" that fit together and are characteristically sculptured (Fig. 6.1*B*). A yellow-brown pigment generally masks the chlorophyll *a,* and this participates in the transferal of radiant energy to the chlorophyll. Food is stored in the form of oil rather than starch. Reproduction is both asexual and sexual.

Numerous representatives from this group play a critical role in the economy of nature, serving as a starting point for aquatic food chains and thus providing the ultimate food source for many fish. Because of the hard and chemically inert silica walls, deposits of fossil diatoms, called *diatomaceous earth,* are mined for industrial use as insulation, as an abrasive, and as a filtering agent.

**Dinoflagellates (Phylum Pyrrophyta).** This group contains motile, photosynthetic, unicellular organisms that have been considered protozoa by zoologists and protophyta by botanists. They have two flagella, one circling the cell and the other trailing in the body axis, both typically occupying grooves in the rigid shell. These marine and, less commonly, freshwater organisms contain yellow-green or yellow-brown plastids, and their food reserve is starch or oil. Although they carry on photosynthesis they also require some complex organic materials, such as vitamin $B_{12}$, in their diet. Reproduction takes place asexually. These organisms can produce toxins that cause extensive destruction of fish; the massive fish kills of the so-called "red tide" are caused by the occurrence of very large and dense dinoflagellate populations.

***Euglena* (Phylum Euglenophyta).** The euglenoids are both motile and photosynthetic, and they have been considered plants by botanists and animals by zoologists, though, as we have suggested, there is no real significance in classifying them in either group. *Euglena* is the genus name for a group of species found in ponds, often forming extensive green patches of surface scum or "blooms."

MORPHOLOGY. *Euglena* is typically spindle-shaped, varying from 25 to more than 100 microns in length (Fig. 6.1*F*). The organism has no cell wall but is surrounded by a stiff *pellicle* (*periplast*). A single whip-like flagellum extends from a "gullet" at the anterior end. (Electron microscopy has revealed a second flagellum.) The basal enlargement of the gullet is called the *reservoir;* into this empty several small contractile vacuoles. A red *stigma,* or "eyespot," is located in the anterior region, near the reservoir. Chloroplasts are ovoid and usually numerous. Granules of *paramylum* (a carbohydrate storage product similar to starch) occur in the cytoplasm. The nucleus is single and centrally located, the chromatin concentrated in its center.

PHYSIOLOGY. Nutrition is either autotrophic or saprophytic, *Euglena* being able to maintain itself in darkness by absorbing dissolved organic material from its surroundings. In the dark, *Euglena* gradually loses its green color. Both pigmented and nonpigmented forms require vitamin $B_{12}$, and other basic nutritional factors may be required by various species. *Euglena* probably does not ingest solid food. Locomotion is accomplished by the flagellum. A form of rhythmic, stationary contraction ("euglenoid movement") occurs occasionally; it is not a form of locomotion.

REPRODUCTION. This occurs asexually, by longitudinal fission, division beginning at the anterior end (Fig. 6.1G). *Euglena* forms *cysts,* in which individuals may survive drying and other unfavorable conditions.

**Protozoa (Phylum Protozoa).** This phylum includes all organisms referred to as one-celled animals. It is usually divided into four classes, though some authorities consider the differences great enough to give these groups phylum rank. The classes are most easily separated on the basis of methods of locomotion. (Because of overlapping characteristics, the first and second groups are sometimes combined.)

(1) MASTIGOPHORA (FLAGELLATA). Locomotion by *flagella,* one to many elongated whiplike structures that function as propellers. Free living and parasitic.

(2) SARCODINA (RHIZOPODA). Locomotion by *pseudopodia* ("false feet"), projections of flowing cytoplasm. Free living and parasitic.

(3) SPOROZOA. Usually passive in movement. All parasitic.

(4) CILIATA. Locomotion by *cilia,* numerous minute, hairlike processes that function as oars. Free living and parasitic.

*Amoeba* and *Paramecium* (belonging to the Classes Sarcodina and Ciliata, respectively), and *Euglena* (Phylum Euglenophyta, but sometimes considered a member of the Class Mastigophora of the Phylum Protozoa) are probably the most frequently studied free-living, unicellular organisms in biology laboratories. The best-known representative of the Class Sporozoa is the organism causing malaria. The following examples illustrate the morphology, physiology, and reproduction of protozoans, but it should be remembered that the group consists of many diverse forms.

AMOEBA. This is a genus of naked protozoa having an irregular and changing body form (Fig. 6.1H). Two other amoeboid forms, both with external shells, *Arcella* and *Difflugia,* are types of Sarcodina that, like *Amoeba,* occur in freshwater ponds. The geologically important Foraminifera and Radiolaria are also shelled Sarcodina. Other representatives of the group are parasitic, one being the cause of a serious form of dysentery.

*Morphology.* An amoeba is approximately 200 to 300 microns in diameter (large enough to be seen with the naked eye). Its external form is asymmetrical, continuously changing during locomotion. The cytoplasm is divided into an external clearer region, the *ectoplasm,* and an inner more opaque region, the *endoplasm.* A single, definite nucleus, shaped like a flattened sphere, is present. There is typically one contractile vacuole, and there are numerous food vacuoles.

*Physiology.* Locomotion takes place by flowing into extensions of cytoplasm, the *pseudopodia.* Solid food is ingested (*holozoic nutrition*) as pseudopodia enclose food vacuoles (Fig. 6.1H). This process is called *phagocytosis.* Digestion takes place in the food vacuoles. There is no definite place at which undigested material is egested. The contractile vacuole excretes excess water which enters because of the higher osmotic pressure inside the cell.

*Reproduction.* This takes place asexually, by fission. A form of *sporulation* (multiple fission) has also been observed.

PLASMODIUM. This is the genus name of the causative organism of malaria, a representative of the Class Sporozoa. It is parasitic in insects and in the blood of vertebrates. For details of its relation to man and the mosquito (its complex life history) see Figure 6.4.

PARAMECIUM. *Paramecium* is a genus of rather large, spindle- or cigar-shaped protozoa that move by means of cilia (Fig. 6.1I). In an infusion of hay in pond water there soon appear various kinds of ciliates as well as *Paramecium.* Other forms commonly observed include *Stentor, Stylonychia, Tetrahymena,* and *Vorticella.*

*Morphology.* This protozoan is about 250 microns long and is asymmetrically cigar-shaped. A spiral *oral groove* leads back from the anterior end to the *gullet,* about halfway back. The ectoplasm is thin, and it contains numerous *trichocysts,* structures which (presumably in defense) are extruded and hardened as threads. One *macronucleus* is present and one or two *micronuclei;* the former controls cellular metabolism, while the latter is involved in cell division. Typically there are two contractile vacuoles and numerous food vacuoles (Fig. 6.1I). The beating of the

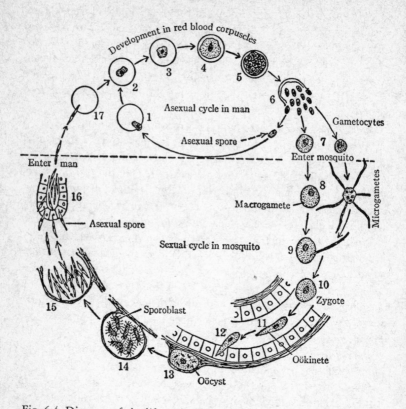

Fig. 6.4. Diagram of the life cycle of the malarial parasite, showing the asexual cycle in man (*1–6, 17*) and the sexual cycle in a mosquito of the genus *Anopheles* (in stomach *8–12,* cyst on stomach wall *13–15,* in salivary gland *16*). (Reprinted by permission from *Animal Biology* by R. H. Wolcott, published by the McGraw-Hill Book Company.)

cilia is coordinated by a neurofibril that connects the basal bodies of the cilia.

*Physiology.* Nutrition is holozoic, bacteria and other small food particles being ingested through the gullet and formed into food vacuoles. These follow a definite course through the cytoplasm during the process of digestion, and the undigested material is egested at a definite region on one side, the *cytopyge.* The contractile vacuoles, as in the amoeba, regulate the osmotic pressure.

*Reproduction.* This may be asexual, by transverse fission, the macronucleus dividing amitotically and the micronucleus or micronuclei dividing mitotically—but within the nuclear membrane (Fig. 6.1*J*). Sexual reproduction takes place by *conjugation*, in which nuclear exchange takes place between two paramecia. These two individuals then separate and each divides rapidly into four small individuals. A nuclear reorganization comparable to this process but involving only a single individual also takes place in *Paramecium*. This is called *endomixis*.

*Part Two*

# MULTICELLULAR ORGANISMS

# Chapter VII

# THE MULTICELLULAR ORGANISM

Multicellular plants and animals have a level of organization within which differentiation occurs among cells or even among groups of similarly functioning cells. Such organisms are generally larger and more complex than unicellular ones, and they constitute conspicuously important population units in many environments. In this chapter we consider general characteristics of multicellular organisms; and in the next six chapters we shall summarize the morphology, physiology, and reproduction of multicellular plants and animals.

## THE ORGANISMAL CONCEPT

Either as a single cell or as a group of cells, the individual organism behaves as a unit. It has organization: the parts are subordinate to the whole, whether they are parts of cells or whole cells. This is the *organismal concept,* one of the most important concepts in biology.

## COLONIES AND ORGANISMS

An organism may exist as a separate individual or as one of an aggregation or group. The organisms in such a group may be unicellular or multicellular. In either case, when organisms of the same kind occur together they form a *colony.* A colony of independent unicellular organisms is comparable to a colony of multicellular ones, not to a single multicellular organism. The individuals may be quite independent, surviving and reproducing when separated; or they may show some division of labor. *Volvox* provides an illustration of a colonial unicellular form with some division of labor. Its spherical colony of individual green flagellates has both nonreproductive and reproductive cells. This division of labor does not make *Volvox* a multicellular organism, however, as it is similar to what occurs in colonies of multicellular organisms that have nonreproductive and reproductive individuals (e.g., ants).

A multicellular organism is an individual, not a colony of independent cells. Cells and cell groups within the organism have varying degrees of dependence upon each other. Even interdependent cells or groups of cells may, however, have the ability to regenerate following various types of injury, or separation from the body proper (Chaps. X and XIII).

# CHARACTERISTICS OF THE MULTICELLULAR ORGANISM

Unicellular and multicellular organisms have fundamental similarities. There are certain basic differences between them, however, primarily related to the differentiation occurring during the development of a multicellular body from a unicellular zygote.

**Similarities with Unicellular Organisms.** Organisms, whether unicellular or multicellular, are the basic units of nature. As such, they all possess, to some degree, internal complexity and division of labor, and they have similar genetic and ecological characteristics.

INTERNAL COMPLEXITY. The idea that unicellular organisms are simple in structure because they are small is incorrect. Many protozoa, for example, are extremely complicated in internal structure, containing many different highly specialized organelles.

DIVISION OF LABOR. This is another characteristic common to one-celled and many-celled organisms. Among the former, the different parts of a single cell illustrate the principle; in multicellular forms division of labor is among the different cells.

GENETIC RELATIONSHIPS. The production of progeny that resemble the parents but that are also potentially different through recombination or mutation (Chap. XVII) is a characteristic of both single-celled and many-celled organisms.

ECOLOGICAL RELATIONSHIPS. Unicellular and multicellular organisms compose populations that have similar characteristics. All require energy and nutrients from the environment (Chap. XX). And all populations, whether of unicellular or multicellular organisms, fit into a limited series of functional positions in the ecosystem (e.g., the unicellular diatoms of the sea and the multicellular grasses of the prairie occupy equivalent functional positions as producers).

**Differences from Unicellular Organisms.** The increased complexity of multicellular organization results in many basic differences that distinguish multicellular from unicellular organisms.

SIZE. In general, unicellular forms are smaller than multicellular ones, but the largest of the former are many times larger than the smallest of the latter.

KINDS OF DIFFERENTIATION. In one-celled organisms, differentiation is between the parts of a single cell; in multicellular organisms it takes the form of differences between individual cells and groups of cells.

*Tissues.* A tissue is a group of cells having the same structure and function. In multicellular animals these are epithelial, connective and supporting, blood (transport), muscular, and nervous tissues. In multicellular plants they are epidermis, parenchyma, sclerenchyma, collenchyma, and vascular tissues (xylem and phloem). (Figs. 7.1 and 7.2.)

*Organs.* An organ is a group of cells or tissues associated together for some common function. It is not necessarily of uniform structure throughout. The following are examples in animals: brain, stomach, foot, fin, wing, biceps, femur. In plants examples are: root, stem, leaf, stamen, carpel.

*Organ Systems.* Organ systems are groups of organs involved in the same functions. The term is not satisfactorily applied to plants, though we do distinguish between vegetative (metabolic) organs and reproductive ones. It is quite common in describing animal structures, e.g., the circulatory, digestive, respiratory, nervous, and locomotor systems.

**Growth and Reproduction.** In unicellular forms growth is associated with increase in cell size. In multicellular plants and animals it involves not only growth in size of individual cells but also increase in the number of cells, through division. Reproduction is a separate process, involving the activity of a specialized group of cells, the germ cells. Since reproduction involves only a few cells, the parent organism generally lives on.

Asexual reproduction occurs in some multicellular organisms but the usual process is sexual. Haploid gametes, which may be alike (*isogametes*) or unlike (*heterogametes*) fuse to form the diploid zygote from which a diploid individual develops by mitosis. The typical life cycle in animals involves haploid cells only as the mature gametes, these formed by meiosis during *gametogenesis*. In most plants, however, the haploid gametes are formed by ordinary mitosis in a plant (*gametophyte*) whose cells are all haploid. The zygote produced in fertilization divides by mitosis, forming a diploid plant (*sporophyte*), which forms reproductive cells called *spores*. These are haploid; they are formed by

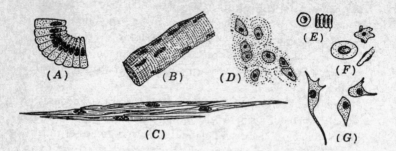

Fig. 7.1. Representative animal tissues. (*A*) Epithelium—cells of the mucosa lining the frog's intestine. (*B*) Striated muscle tissue from the leg of a grasshopper. (*C*) Smooth muscle tissue from the frog's intestine. (*D*) Cartilage (supporting tissue) from the frog. (*E*) Human red blood corpuscles. (*F*) Blood corpuscles of the frog. (*G*) Nerve cells from the spinal cord of a rabbit.

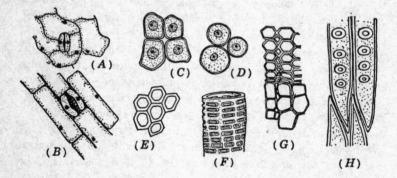

Fig. 7.2. Representative plant tissues. (*A*) Epidermis from leaf of turnip. (*B*) Epidermis from leaf of corn. (*C*) Wood parenchyma cells from stem of *Aristolochia* (Dutchman's pipe). (*D*) Parenchyma cells from cortex of corn root. (*E*) Sclerenchyma fibers (in cross section) from pericycle of stem of *Clematis*. (*F*) Pitted vessel from xylem (vascular tissue) of sycamore tree. (*G*) Cross section of tracheids (vascular tissue) from white pine—two adjacent annual rings: summer wood above, spring wood of the succeeding year below. (*H*) Longitudinal section of tracheids with bordered pits, from white pine.

meiosis during *sporogenesis*. Each spore can give rise to a new haploid plant (gametophyte). Thus, in most plants, the reproductive cycle involves two generations that alternate, a haploid and a diploid one, while in animals there is no haploid generation.

# THE ORIGIN OF MULTICELLULAR ORGANISMS

A variety of untestable speculations have been developed about the evolution of various multicellular organisms—simply by analyzing the various existing levels of organization. Multicellular plants are believed to have developed through cohesion of daughter cells. Various major algal and fungal groups are considered to have developed independently, and the bryophytes and tracheophytes are believed to have further evolved as separate groups from branched green algal filaments. The origin of multicellular animals is considered by some to have developed also as a result of cohesion of dividing unicellular forms. Sponges probably had a separate origin from other multicellular animals (*Metazoa*), whereas the existing flatworms and coelenterates, considered primitive metazoans, reflect a common ancestor that could have developed from an advanced colonial form. In contrast to this theory some biologists have suggested that a single cell having many nuclei (*syncytium*) might have preceded compartmentalization into multicellular units. Although some simple syncytial flatworms appear to represent the intermediate condition, their gametes are cellular and development proceeds through normal mitosis. There is also a highly speculative theory that certain one-celled animals may have developed through a loss of the multicellular condition.

# Chapter VIII

# THE MORPHOLOGY OF MULTICELLULAR PLANTS

We here consider as multicellular plants representatives of not only the phyla invariably treated as members of the plant kingdom but also some that may be placed in the Protista. These organisms can be divided into two major groups, the Thallophyta and the Embryophyta. (In other classifications four divisions are recognized, the Thallophyta, Bryophyta, Pteridophyta, and Spermatophyta. The Embryophyta include the last three of these divisions.)

The Thallophyta are an artificial assemblage of plants without the specialized organs (leaves, stems, roots) characteristic of higher plants. Their body is called a *thallus*. Thallophytes are of two distinct types for which the common names *algae* and *fungi* are used; the algae have chlorophyll and are autotrophic, the fungi are heterotrophic. The Embryophyta are a logical assemblage of plants comprising two phyla, the Bryophyta and the Tracheophyta, related by the possession of a protected *embryo*.

## THE PRINCIPAL CRITERIA FOR CLASSIFICATION OF PLANTS INTO PHYLA

The phyla (or divisions) are distinguished from each other by combinations of characteristics, not by single sets of characteristics.

**Body Organization.** The level of organization often, but not always, differentiates groups. This level may be cellular, multicellular, tissue system, or organ system, and within these there are further distinctions. If cellular, the contrast between procaryotic and eucaryotic structure is of basic importance. Cell groupings may involve filaments, sheets, or more complex tissues. The body of the plant may have distinctive functional and structural units such as roots, conducting tissue, supporting tissue, and leaves. The types of flagella of cells such as gametes may be significant—positions, numbers, and structure.

**Reproductive Characteristics.** The various characteristics of plant life cycles, and the relative duration and structural prominence of sporophyte and gametophyte generations are of great importance in differentiating major groups.

**Metabolic Differences.** The mode of nutrition as reflected by the presence or absence of chlorophyll, and the specific types of chlorophylls and accessory pigments, are criteria used in plant classification. Differences in metabolic pathways are reflected in the various types of food reserves and other components that can be manufactured by certain organisms but not by others. In the final analysis, the presence or absence of specific enzymes, as well as enzyme and isozyme structure, reflects true differences in metabolic capabilities. These differences provide criteria for classification into major groups.

**Broad Distribution and Habitat Preferences.** Although specific habitat preferences are reflected by adaptations in morphology and physiology, these adaptations have sometimes resulted in rather stringent requirements. As a result, the well-informed biologist knows what particular types of plants to expect in major habitat types.

## MORPHOLOGY OF MULTICELLULAR FUNGI

Fungi have no chlorophyll and are essentially heterotrophic, depending upon the products of other organisms. There are parasitic types that cause various plant or animal diseases, and types that are saprophytic, living on decomposing plant or animal material. Fungi vary in size from small microscopic organisms to moderately large mushrooms and bracket fungi. Varied, sometimes complex, reproductive cycles occur in the group. Bacteria (formerly considered fungi), yeasts, and slime molds were all described in Chapter VI. Here we shall consider representatives of only one phylum, the Eumycophyta, or true fungi. (Some fungi are placed in other, small phyla.) The phylum Eumycophyta is often divided into three classes, the Phycomycetes, Ascomycetes, and Basidiomycetes. Some authorities recognize the Fungi Imperfecti (those forms in which key reproductive stages are lost or have never been observed), and some accept additional classes of the Eumycophyta.

The fungal thallus is differentiated into filamentous vegetative or food-gathering portions and reproductive structures. Fungal filaments, *hyphae* (sing. hypha), grow rapidly from their tips, usually forming branched mats called *mycelia* (sing. mycelium). The hyphae secrete enzymes into their surroundings, and these break down complex organic compounds into the subunits that can be taken into the cell. Fungi generally also require some complex growth factors such as vitamins. Food reserves include glycogen and oil droplets. The diverse fungal reproductive structures are used in the identification and classification of species. Fungi typically produce numerous unicellular spores. The motile spores of Phycomycetes, with a variety of flagellar types, are called *zoospores.* Spores produced by asexual reproduction are called *mitospores.* Sexual reproduction generally produces a diploid spore (zygospore) that is thick-walled and dormant, enabling the organism to survive times of unfavorable climatic conditions. Potato blight (which caused the Irish famine), downy mildew of grapes, Dutch elm disease, "ringworm," athlete's foot, and smut and rust plant diseases are caused by fungi. Various ascomycete genera (e.g., *Penicillium*) are alternately enemies of man, causing food spoilage, and benefactors, as sources of *antibiotics,* which inhibit the growth of pathogenic organisms. Fungi also play an important role in the decay of organic material and the resulting recycling of nutrients in nature. For reproduction in the fungi see Chapter X.

**Black Bread Mold (Class Phycomycetes).** (Fig. 8.1.) The mycelium extends down into the moist bread as food-absorbing hyphae

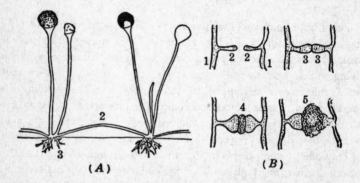

Fig. 8.1. *Rhizopus* (black bread mold). (*A*) Portion of *Rhizopus* mycelium. The vertical hyphae are sporangiophores, each bearing at its apex a sporangium that bears asexual spores. The horizontal hyphae (2) are called stolons, and those (3) extending into the substrate are rhizoids. (*B*) Stages in sexual reproduction involve two different hyphae (1). Steps in the formation of a zygote are shown in the details numbered 2–5: (2) progametes, (3) isogametes, (4) fertilization, (5) zygote. (Reprinted by permission from *General Botany* by Harry J. Fuller and Donald D. Ritchie, published by Barnes & Noble, Inc.)

(*rhizoids*) or extends upward as spore-bearing hyphae (*sporangio-phores*). The hyphae lack cell walls and are multinucleate. The genus name for the black bread mold is *Rhizopus.*

**Red Bread Mold (Class Ascomycetes).** The mycelium of this tropical and subtropical mold is made up of hyphae with incomplete cross walls, which allow cytoplasmic exchange. Sexual union may result in cells with two unfused nuclei, one from each mating type, and this *dicaryotic* condition may be perpetuated by the individual mitosis of each nucleus. The *ascospores,* characteristic of this class, are produced in an elongated sac, an *ascus.* The spores occur in an ascus in groups of eight, following nuclear fusion, meiosis, and one mitotic division. This mold, whose genus name is *Neurospora,* is used extensively in experimental studies in genetices.

**Mushrooms (Class Basidiomycetes).** (Fig. 8.2.) A filamentous mycelium forms an underground portion. It consists of interlacing hyphae, in which cross walls are present. In some species the mycelium

Fig. 8.2. A mushroom, part of cap removed to expose gills.

forms, through specific morphogenetic processes, the commonly observed fleshy fruiting body or basidiocarp. These fruiting bodies may consist of spheres, such as puffballs, or tough parasols, such as mushrooms. The basidiocarp consists of a *stalk* (stipe) and a *cap.* On the underside of the cap are borne thin plates, the *gills,* which radiate outward from the stalk. Extending out at right angles to the surface of the gills are sterile hyphae and swollen cells known as *basidia,* each of which characteristically forms four external spores in cellular extensions. In Basidiomycetes the dicaryotic condition is of increased relative length and there is a reduced capacity for asexual reproduction.

## MORPHOLOGY OF MULTICELLULAR ALGAE

Algae have chlorophyll and are, therefore, autotrophic. In many cases they contain other pigments as well. In color they appear green, red, golden, or brown, depending upon the supplementary pigments that are present. Algae vary in size from microscopic species to large seaweeds that may be more than one hundred feet long. Some are relatively complex in vegetative structure, and many have complex reproductive cycles. They are common in the sea and in fresh water, but on land they are limited to damp locations. Certain phyla of algae, the Cyanophyta, Chrysophyta, Pyrrophyta, and Euglenophyta were described in Chapter VI. These will not be included here as they have no true multicellular representatives, but we shall consider in this chapter three phyla that do have multicellular members, specifically the Chlorophyta (green algae), Phaeophyta (brown algae), and Rhodophyta (red algae).

**Green Algae (Phylum Chlorophyta).** This is a large group, including many species of unicellular and multicellular organisms. (Unicellular forms were described in Chapter VI.) The green algae have chlorophylls *a* and *b,* and the reserve food is starch stored in chloroplasts. These characteristics, as well as the cellulose cell walls and whiplash flagella (when flagella are present), are shared with the higher plants. For this reason some classification schemes combine green algae and the embryophytes. There are green algae with life cycles that include sexual and asexual reproduction and alternation of haploid and diploid generations. Most species live in fresh water, but some are marine, and others are found on land in moist habitats. For the reproduction of the following representatives of multicellular green algae, see Chapter X.

ULOTHRIX. (Fig. 8.3*A*.) This freshwater alga consists of an unbranched filament of cells in linear order, each cell with a single chloroplast, the latter shaped like a broad transverse belt. The chloroplast contains one or more small centers, *pyrenoids,* which are surrounded by starch. The filament is attached to the substrate by a modified basal cell, the *holdfast.*

ULVA. *Ulva* is a marine alga called sea lettuce. It is common between low and high tides on both coasts. The plant is a sheetlike thallus, two cells thick—the cells elongated at right angles to the surface. Rhizoids from some of the lower cells form a holdfast.

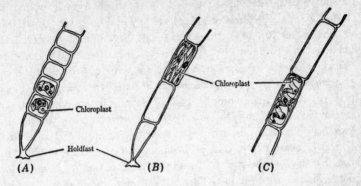

Fig. 8.3. Green algae. Parts of filaments showing vegetative cells of (*A*) *Ulothrix*, (*B*) *Oedogonium*, (*C*) *Spirogyra*. Cellular details are shown in only one or two cells of each filament. The dark ovoid bodies are the nuclei, the lighter ones, the pyrenoids.

OEDOGONIUM. (Fig. 8.3*B*.) A filamentous, freshwater alga, this is similar in appearance to *Ulothrix*. The chloroplast is a cylindrical network with many pyrenoids.

SPIROGYRA. (Fig. 8.3*C*.) This freshwater form is not attached. It forms unbranched filaments of elongated cells in which the chlorophyll is in one or more spiral chloroplasts, each with numerous pyrenoids. The nucleus is suspended in the center of the cell in cytoplasmic strands.

**Brown Algae (Phylum Phaeophyta).** The brown algae are all multicellular organisms, ranging from simple branched filaments with little cell specialization to giant seaweeds. The broad ribbonlike thallus of the kelp (*Laminaria*) may be over 100 feet in length. Simple differentiation into a rootlike *holdfast*, a stemlike *stipe,* and a leaflike *blade* occurs. In the rockweed (*Fucus*), air bladders (*floats*) are also present. Specialized reproductive organs are found in this group, and there are a variety of life cycle types that include asexual and sexual reproduction. The gametes are flagellated. The photosynthetic cells contain chlorophylls *a* and *c* and the brown accessory pigment, *fucoxanthin.* Excess photosynthate is stored in the form of lipids and *laminarin,* a soluble sugar. The cellulose cell wall is surrounded by a gelatinous layer that contains *alginates.* Brown algae have been used as fertilizers, a source of iodine, and a source of alginates—which are used as food stabilizers in ice creams, salad dressings, and candy bars, and as components of underwater paints, some plastics, and waterproof cloth. Brown algae are generally found attached to the ocean floor within intertidal regions, and, except for *Sargassum,* do not survive when suspended freely in the water.

**Red Algae (Phylum Rhodophyta).** These relatively small plants (a few inches to two or three feet long) have one or more elaborate filaments that form an axis from which lateral, often feathery, filaments branch. No cells are flagellated. There are numerous complex life cycle patterns in this group, with asexual and sexual reproduction and examples of either free-living or parasitic sporophyte and gametophyte stages. Chlorophyll *a* is the main photosynthetic pigment, but accessory pigments, including the red *phycoerythrin,* enable the red algae to live at greater ocean depths than other photosynthetic plants, utilizing light rays that have not been filtered out by the surface water. *Floridean starch* is the food storage product. The cell wall of cellulose is surrounded by a mucilaginous layer, which is an economically important source of agar. These algae, which are mainly marine, are most commonly found along rocky seacoasts in the lower intertidal zone or in deeper water, but some freshwater species occur in cold swift-flowing streams. Some species found in coral reefs are believed to be as important in reef building as are coral animals.

## MORPHOLOGY OF THE EMBRYOPHYTA

In the Embryophyta the fertilized egg remains in the female sex organ during early stages of development as an *embryo.* Furthermore, both male and female sex organs are multicellular, having enclosed layers of nonreproductive or *sterile cells.* The most primitive embryophytes have a body that is essentially a thallus, but stems, leaves, and primitive-type roots appear early in the evolutionary sequence. The next advances involve the differentiation of vascular or conducting tissue, and this made possible the evolution of large land plants. Within the Embryophyta there is a continuous alternation of the so-called sporophyte and gametophyte generations. The sporophyte plant, which is diploid, develops from a zygote resulting from fused gametes. The gametophyte plant, which is haploid, develops from a haploid spore produced by meiosis in the sporophyte. The gametophyte produces gametes. Although the two types of plants alternate in the life history of every higher plant, the sporophyte generation becomes progressively more prominent and the gametophyte generation less so.

**Mosses and Liverworts (Phylum Bryophyta).** This group comprises small green plants with stemlike and leaflike structures and rhizoids that anchor them in the soil. The life cycle involves alternation of generations with the sporophyte parasitic on the gametophyte. A few plants of this phylum are aquatic, but most of them are terrestrial, active only in damp environments.

Moss. (Fig. 8.4.)

*Vegetative Parts.* True roots are absent, but rhizoids function as roots. There is an erect "stem" with scalelike "leaves." Cells of the stem have little differentiation; outer ones contain chloroplasts, inner ones

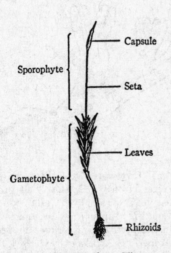

Fig. 8.4. A moss plant. The sporophyte is parasitic on the gametophyte.

form a rudimentary conducting tissue. True vascular tissue does not occur in bryophytes. The leaves are thin and narrow, and are usually only one cell layer thick. The cells of leaves contain chloroplasts. These vegetative structures occur only in the gametophyte generation.

*Reproductive Parts.* (See Chap. X for the life cycle.) At the upper end of the stem of the gametophyte there is either a sperm-producing organ (*antheridium*) or an egg-producing organ (*archegonium*) or both. The mature sperms and eggs are known collectively as *gametes.* From the fertilized egg (ovum), which remains in the archegonium, there grows out a long stalk (*seta*) bearing at its upper end the *capsule.* Spores are formed in the capsule. The seta and capsule, along with the *foot,* which is embedded in the gametophyte, constitute the sporophyte.

The cells of the sporophyte lack chlorophyll and are metabolically dependent upon the gametophyte.

LIVERWORTS. (Fig. 8.5.) Some liverworts have only a thallus body, with rhizoids. Others have a "stem" and "leaves." In contrast to mosses, however, they have typically a prostrate rather than an erect form. Antheridia and archegonia are present. The sporophyte generation is relatively less conspicuous than in the mosses.

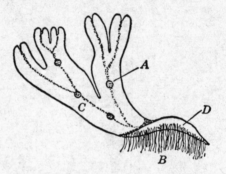

Fig. 8.5. Liverwort. Except for the small structures labeled (*A*), this is all gametophyte. (*A*) Sporophyte. (*B*) Rhizoids. (*C*) and (*D*), upper and lower surfaces of thallus, respectively. (Reprinted by permission from *General Botany* by Harry J. Fuller and Donald D. Ritchie, published by Barnes & Noble, Inc.)

**Vascular Plants (Phylum Tracheophyta).** These are small to large, chlorophyll-bearing plants with distinct roots, stem, and leaves. All have vascular tissue in a cylinder (*stele*) with both xylem and phloem. The sporophyte is the dominant stage in the life cycle. In primitive Tracheophyta the gametophyte may be metabolically independent, but in most plants in this group the gametophyte is parasitic on the sporophyte. Plants of this phylum dominate land habitats; a few species, however, are aquatic.

The Phylum Tracheophyta is divided into four subphyla, the Psilopsida (primitive, mostly extinct forms), Lycopsida (club mosses), Sphenopsida (horsetails), and Pteropsida (ferns and seed plants). The subphylum Pteropsida is divided into three classes, the Filicineae (ferns), Gymnospermae (pine, etc.), and Angiospermae (flowering plants). The angiosperms and gymnosperms have also been combined into a single group called the spermatophytes, the remaining groups, including the ferns, combined into the pteridophytes.

Life histories of representatives are summarized in Chapter X; Chapter IX is devoted largely to the physiology of angiosperms.

CLUB MOSSES (SUBPHYLUM LYCOPSIDA). The sporophyte is moss-like, the stem bearing many small leaves that are spirally arranged. Reproductive leaves, called *sporophylls,* bear sporangia. In some species a cluster of sporophylls forms a *strobilus* at the tip of a stem. Club mosses are abundant in the tropics; they were dominant plants in the Carboniferous period, when many of them were tall and treelike.

HORSETAILS OR SCOURING RUSHES (SUBPHYLUM SPHENOPSIDA). The rhizome branches into jointed stems that bear rudimentary leaves in whorls at the joints. Some branches are vegetative in function; others are reproductive, having conelike *strobili* at the tips. The strobilus consists of an axis bearing shield-shaped *sporangiophores,* which bear sporangia. Although living forms are small, giant horsetails existed in the Carboniferous period. Because the epidermis of the stems contains silica, "scouring rushes" were used in pioneer days for cleaning pots and pans.

FERNS (SUBPHYLUM PTEROPSIDA, CLASS FILICINEAE). The members of this class have an alternation of generations in which both gametophyte and sporophyte are independent; they develop large leaves; and they lack seeds. (Fig. 8.6.) With club mosses and horsetails they formed the dominant vegetation of the Carboniferous period.

*Vegetative Parts.* The gametophyte is a small, heart-shaped plant, the *prothallus,* bearing rhizoids. Its cells contain chloroplasts. The sporophyte is the plant we think of as a fern, consisting of *fronds* (leaves) with central stalk and lateral leaflets, growing from a large underground stem (*rhizome*). The latter is anchored by small roots which have root hairs. The cells of these roots are differentiated into various kinds of tissues: *epidermis,* the outer layer of cells from some of which the root hairs elongate; *parenchyma,* simple thin-walled cells; and *sclerenchyma,* thick-walled supporting cells. Vascular tissue occurs, usually in the form of "bundles," each consisting of a core of *xylem,* large hollow woody cells, surrounded by strands of *phloem,* small cells usually containing protoplasm. The *meristem* (growing part of the root tip) is a single cell. A *rootcap* is present. Epidermis, sclerenchyma, parenchyma, and vascular tissue are also present in the rhizome, and there is a *cortex* of sclerenchyma immediately under the epidermis. Internal bands of sclerenchyma are present. The xylem and phloem are com-

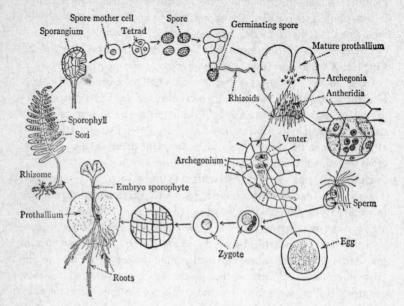

Fig. 8.6. Diagrams showing stages in the life cycle of a true fern (*Polypodium*). (Reprinted by permission from *A Textbook of General Botany* by R. M. Holman and W. W. Robbins, published by John Wiley & Sons, Inc.)

bined in one *fibrovascular bundle,* the former usually surrounded by the latter. The leaf has an epidermis containing *stomata* (pores whose size is regulated by their marginal cells, the *guard cells*). Inner sclerenchyma is absent. The vascular bundles are branched to form *veins.* Chloroplasts are present in the nearly uniform parenchyma cells that occupy most of the leaf, and these constitute the *mesophyll.*

*Reproductive Parts.* Sperm cells are produced in antheridia, small spherical bodies on the underside of the prothallus. A single ovum develops in each of the archegonia, which are flask-shaped bodies also on the underside of the prothallus. Each sperm cell is spirally twisted, and motile through the action of many long flagella. Spores develop in *sporangia.* The latter occur in groups called *sori* (sing., *sorus*), brownish or yellowish, flattened ovoid structures on the under surface of the leaflets.

PINE (SUBPHYLUM PTEROPSIDA, CLASS GYMNOSPERMAE). Gymnosperms are mostly large trees, this conspicuous stage being the sporophyte generation. The gametophyte is inconspicuous, and it is dependent upon the sporophyte. *Seeds* are produced; they are borne naked on megasporophylls (Fig. 8.8).

The best-known gymnosperms are the conifers (Order Coniferales), which include such familiar trees as the pines, hemlocks, firs, spruces, cedars, and redwoods. Most conifers are evergreens with needlelike or scalelike leaves. They are abundant plants, particularly in the colder regions of the world, and they are of great economic importance. The pine has been selected here as an example of a gymnosperm.

Other living orders of gymnosperms are the Cycadales (cycads) and the Ginkgoales (ginkgo or maidenhair tree). The extinct seed ferns (Order Pteridospermales) resembled ferns in their vegetative features but produced seeds.

*General Characteristics.* A pine has a straight axial stem (*trunk*) with lateral branches, the trunk continued downward in a *taproot* with side roots. The leaves ("needles") are borne on dwarf shoots, in clusters or bundles, the number of needles in a bundle being characteristic of a species. Reproductive structures are confined to *cones* (strobili) of two kinds—ovule-producing and pollen-producing. The seeds mature in the former. The embryo contains several *cotyledons* (embryonic leaves).

*Vegetative Parts.* The roots and trunk and their branches attain considerable size, by secondary thickening. Xylem and phloem are produced on the two sides of a *cambium* layer (meristem tissue), phloem toward the outside, xylem inside. Annual rings are produced in both trunks and roots. Dwarf shoots occur on the trunk and branches. The leaves are elongated, needlelike structures. The epidermis of the leaves contains deeply sunken stomata and is underlaid by sclerenchyma. The vascular region (*stele*) of the leaf is surrounded by a sheath and contains two vascular bundles, each containing xylem and phloem.

*Reproductive Parts.* (Fig. 8.7.) The vegetative parts of the plant are limited to the sporophyte generation. The reproductive structures are cones of two types, borne on the sporophyte. The *staminate* (male) are smaller than the *ovulate* (female) cones. The staminate cones are made up of small scales, *microsporophylls,* each bearing two sporangia on the inner surface. In each sporangium are produced many *microspores,* which germinate into *pollen grains.* The large cones, the ones commonly called "pine cones," are the ovulate cones. These are made up of large scales, *megasporophylls,* each bearing on its inner surface two *ovules,* structures that develop, after fertilization, into seeds.

**(A)**                        **(C)**

**(B)**

Fig. 8.7. Cones of the ponderosa pine. (*A*) Staminate cone, natural size. (*B*) Ovulate cone, one-half natural size. (*C*) Single scale from the latter, bearing two
winged seeds.

FLOWERING PLANTS (SUBPHYLUM PTEROPSIDA, CLASS ANGIO-
SPERMAE). Members of the Class Angiospermae have a dominant,
independent sporophyte generation (the "plant" of common terminology) and a small, dependent gametophyte generation, which is limited
to a few cell divisions within floral structures. *Flowers,* which are
characteristic of this class, are differentiated into *sepals, petals, stamens,*
and *carpels*. Seeds are produced; these are enclosed within carpels
(Fig. 8.8).

The descriptions given here are not based on one particular example
but apply to flowering plants in general. The major range of variations
within the group is covered in this account.

*General Characteristics.* The sporophyte plant is differentiated into
roots, stems, and leaves, which are vegetative in function, and flowers,
which are reproductive. The gametophyte plants develop within the
stamens and carpels of the flower.

The class is divided into two subclasses, the Dicotyledoneae and the
Monocotyledoneae. As the names suggest, members of the former have
two cotyledons (seed leaves) in the embryo, members of the latter but
one (Fig. 8.9). The leaves of dicots have net venation (veins forming
a network), those of monocots have parallel venation. Flower parts
usually occur in fours or fives, or in multiples thereof, in dicots; in
threes or multiples of three in monocots. The stem structure differs, too,
in the arrangement of vascular tissue; in dicots the vascular bundles
form a circle or cylinder, but in monocots they are scattered throughout
the stem.

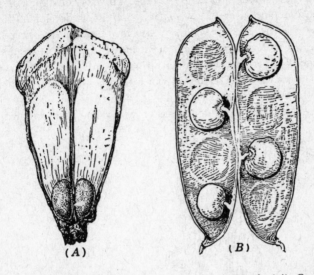

Fig. 8.8. Gymnosperm and angiosperm compared. (*A*) Cone
scale of sugar pine bearing two *naked seeds*. (*B*) Pod of lima
bean open to show the *enclosed seeds*. (Reprinted by permission
from *Fundamentals of Biology* by A. W. Haupt, published by
the McGraw-Hill Book Company.)

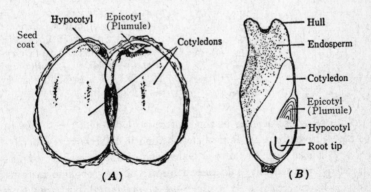

Fig. 8.9. Dicotyledonous (lima bean) and monocotyledonous (corn) seeds
compared. Note: the absence of endosperm in the bean seed is not a characteristic of all dicotyledonous seeds. (Reprinted by permission from
*Fundamentals of Biology* by A. W. Haupt, published by the McGraw-Hill
Book Company.)

*Roots.* The first root to develop from the embryo is called the *primary root,* and its branches are the *secondary roots.* Roots developing from leaves or branches are *adventitious;* they may develop into props or buttresses. A root growing straight down in the stem axis is a taproot. If enlarged for food storage (as in carrots or turnips), it is a *crown root.* Roots with swollen regions in which food is stored (for example, sweet potatoes) are *tuberous roots.*

If we examine the longitudinal section of a young root (Fig. 8.10),

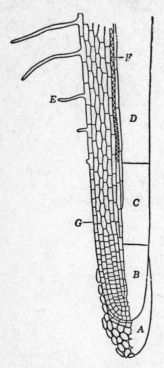

Fig. 8.10. Longitudinal section of young root. (*A*) Rootcap. (*B*) Meristematic zone. (*C*) Zone of elongation. (*D*) Zone of maturation. (*E*) Roothair. (*F*) Stele. (*G*) Epidermis. (Reprinted by permission from *General Botany* by Harry J. Fuller and Donald D. Ritchie, published by Barnes & Noble, Inc.)

we can recognize a series of zones of cells of different types. The tip is covered by a *rootcap.* Immediately behind it is the *meristem* or zone of cell division, and back of that is the *zone of elongation.* The elongating cells are gradually transformed further up the root into maturing cells. The epidermal cells in the *zone of maturation* bear root hairs, which are extensions from single cells. If we examine a cross section (Fig. 8.11*A, B*), we see that the outer layer of cells of the root, the *epidermis,* surrounds a wide *cortex,* the inner layer of which is differentiated into the *endodermis.* The endodermis surrounds the *stele* or cylinder of vascular tissue, consisting of an outer *pericycle* and alternate

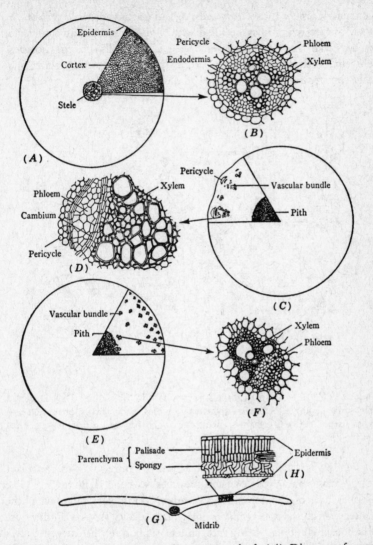

Fig. 8.11. Microscopic structure of root, stems, leaf. (*A*) Diagram of cross section of root of spiderwort (*Tradescantia*), with details shown in sector. (*B*) Stele of same section enlarged. (*C*) Diagram of cross section of dicotyledonous (*Clematis*) stem; the pith extends out to and between the vascular bundles. (*D*) Single vascular bundle enlarged. (*E*) Diagram of cross section of monocotyledonous (corn) stem; the pith (parenchyma of the stele, properly) extends around the vascular bundles. (*F*) Single vascular bundle enlarged. (*G*) Diagram of vertical section through leaf (privet). (*H*) Small section enlarged, part of a vein at the right.

columns of *xylem* and *phloem* cells. Xylem may occupy the center of the root, or that region may be filled with a *pith* of parenchyma cells. In woody plants a cambium layer (of meristem cells) occurs between phloem and xylem, resulting in secondary thickening. In temperate and cold climates *annual rings* may be laid down.

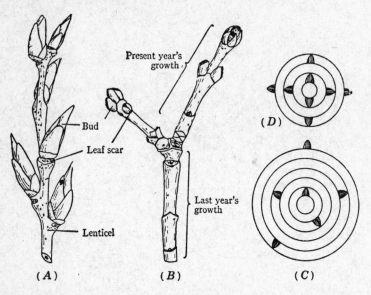

Fig. 8.12. Winter twigs. (*A*) Cottonwood. (*B*) Box elder. (*C*) Diagram showing the angles between adjacent bud scales, buds, leaves, or branches in cottonwood. (*D*) Diagram showing the angles between the same structures in the box elder.

*Stems.* The main stem may constitute a central axis of the plant or it may break up into many branches with no definite axis. The former type of development is *excurrent* (examples: pine, oak), the latter, *deliquescent* (example: elm). Some plants are erect, others are trailing or climbing—being incapable of supporting themselves. Leaves are attached along the stem at intervals characteristic for each particular species. Points of attachment are *nodes,* the stem between two nodes constituting an *internode.* The leaves may be *opposite* each other in pairs, or they may occur in *whorls* of three or more at a node; if they are single at each node, the arrangement is *alternate* or *spiral.* In the latter case, the angle between two adjacent leaves is characteristic for the species. (Fig. 8.12*C, D.*)

At the tip of each stem, and in the *axils* of the leaves (the upper angles between leaves and stem), occur meristem cells. These are enclosed in leafy outgrowths, the *buds.* The buds have the same arrangement on the stem as do the leaves. In trees and shrubs of colder regions, the outer parts of the buds are scalelike and protect the overwintering bud. When the scales drop off in the spring, scars are formed in the bark. These enable one to determine the age of a twig. For examples of winter twigs, see Figure 8.12.

Two distinct types of stem structure occur in the spermatophytes: the dicotyledonous and the monocotyledonous. (Fig. 8.11*C–F.*) In the gymnosperms and dicotyledonous plants, the vascular tissue is concentrated in a cylindrical sheath around the central *pith.* The xylem is internal to the phloem, the cambium lying between them. In woody plants the xylem and phloem are continuous around the periphery; in herbaceous plants, they are usually concentrated in vascular bundles. Outside the phloem lies the cortex, consisting of parenchyma cells and epidermis, and, in some cases, *cork.* The cork is interrupted by holes through it, the *lenticels,* by which the cells of the cortex are in contact with the atmosphere. In the monocotyledonous plants, the vascular tissue is scattered in bundles throughout the stem. There is no central pith. Growth occurs between the vascular bundles, hence they are gradually pushed further from each other.

*Leaves.* Leaves are typically broad, flattened structures attached to a branch by a *petiole* which divides in the leaf *blade* into *veins.* (In contrast with roots and stems, which have *indeterminate* or continuous *growth,* leaves and flowers have *determinate growth,* growth to a definite size.) If it has but one blade, the leaf is *simple;* if several blades, it is *compound.* In the latter, the separate blades are *leaflets.* A single prominent axial vein is usually present, the *midrib.* If the principal veins arise from the sides of the midrib, the leaf is *pinnately* veined. If several prominent veins radiate from the petiole, the leaf is *palmately* veined. In both these cases, minute cross veins form a network, and the leaf has *net venation.* If there are numerous parallel veins running lengthwise of the blade, the venation is *parallel.* Simple leaves are of a great variety of shapes—e.g., linear, elliptical, heart-shaped, with dentate, crenate, or incised margins, etc. (See Fig. 8.13.) In compound leaves, the leaflets may be pinnately or palmately arranged. Leaves may be highly modified—to form tendrils, spines, etc.

The vascular tissue, in bundles, is contained in the petiole and veins. The blade is covered above and below with a layer of *epidermis,* containing at intervals the *stomata*. These consist of pores whose size is regulated by *guard cells*, a pair at each pore. The mid-portion of the leaf, the *mesophyll*, consists typically of one or two layers of vertically

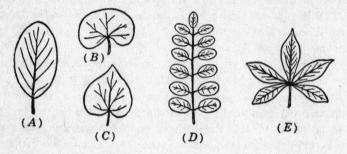

Fig. 8.13. Shapes of leaves. (*A-C*) Simple leaves: (*A*) elliptical, (*B*) kidney-shaped, (*C*) heart-shaped. (*D-E*) Compound leaves: (*D*) pinnately compound, (*E*) palmately compound.

elongated cells under the upper epidermis, the *palisade parenchyma,* and the irregular *spongy parenchyma* occupying the rest of the space. (Fig. 8.11*G, H.*) The mesophyll cells contain the chloroplasts.

*Flowers.* A typical flower consists of four kinds of structures in concentric whorls, attached to the enlarged apex of the stalk (*receptacle*). (See Figs. 8.14 and 8.15.) The structures of the two outer whorls are usually leaflike in form. Various degrees of reduction and fusion occur in either or both whorls. The outer whorl is the *calyx;* its separate elements are *sepals.* They are commonly green and scalelike. The inner whorl is the *corolla;* its separate elements are *petals.* They are usually delicately thin and are often brightly colored. The calyx and corolla together constitute the *perianth.* If radially symmetrical, the flower is said to be *regular;* if it is bilaterally symmetrical, it is *irregular.* The third whorl (from outside in) is made up of *stamens* and is called the *androecium.* Each stamen or *microsporophyll* consists of a basal stalk, the *filament,* and a sac in which the pollen grains form, the *anther.* The inner whorl, the *gynoecium,* consists of *carpels* (*megasporophylls*). In most flowers, the carpels have fused into a single central structure, the *pistil.* It consists of a basal expanded portion, the *ovary;* its vertical extension, the *style;* and the *stigma,* the specialized surface which receives the pollen grains. If the ovary has but a single cavity, it is "single-celled"; if the cavity is divided by one or more partitions, it has several "cells." Several separate pistils are present in some flowers. The mature ovary constitutes the *fruit.*

Various types of floral variations occur. Three variations are related to the positions of attachment of floral structures to the receptacle. If the ovary is superior to (above) the bases of other floral parts, as in the bean and buttercup, the flower is *hypogynous. Perigyny* occurs if (as in the cherry) the ovary is surrounded by a cup upon the edge of which are borne the other floral parts. If the ovary is completely inferior to (below) the other floral parts, the flower is *epigynous.* The flower of the evening primrose is an example of the last type.

Not all flowers have all typical parts. If stamens and pistils are both present in the same flower, it is a *perfect flower;* and, if calyx and corolla are also both present, the flower is *complete* as well as perfect. If stamens and carpels occur in different flowers, the flowers are called *imperfect.* Imperfect flowers—producing gametes of opposite sex—may occur on the same plant, in which case the plant is said to be *monoecious.* If they occur on different plants, the condition is *dioecious.* Corn is monoecious; the cottonwood tree is dioecious.

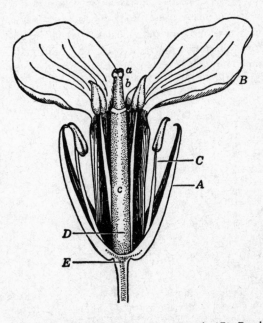

Fig. 8.14. A complete flower. (*A*) Sepal. (*B*) Petal. (*C*) Stamen (anther and filament). (*D*) Pistil: *a,* stigma, *b,* style, *c,* ovary. *E,* Receptacle. (Reprinted by permission from *General Botany* by Harry J. Fuller and Donald D. Ritchie, published by Barnes & Noble, Inc.

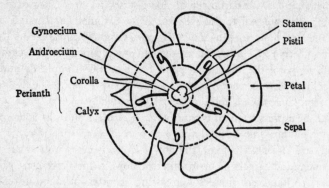

Fig. 8.15. Diagram illustrating the arrangement of parts in a typical flower.

Flowers do not necessarily occur as single structures at the end of a stem. They exist in many complex combinations called *inflorescences.* The pattern of occurrence of individual flowers in the major types of inflorescences is illustrated in Figure 8.16. These fall into two groups: *indeterminate inflorescences,* in which the terminal flower is the youngest

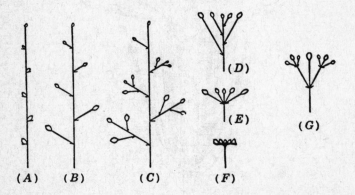

Fig. 8.16. Types of inflorescences. The ovals represent flowers: the larger ovals, the older flowers. (*A*) Spike. (*B*) Raceme. (*C*) Panicle. (*D*) Corymb. (*E*) Umble. (*F*) Head. (*G*) Cyme. (*A-F*) Indeterminate inflorescences. (*G*) Determinate inflorescence. (Modified from various authors.)

(most recently formed), and *determinate* ones, in which the terminal flower is the oldest or first. Among inflorescences illustrated in the figure are the following indeterminate ones: *spike, raceme, panicle, corymb, umbel, head.* The *cyme* is a determinate inflorescence, as is also, of course, a single flower at the end of a single stem.

*Fruits.* A fruit is usually defined as a ripened ovary. It is that organ of the plant that ordinarily contains seeds. A fruit may be derived from one ovary or many, and it may include other structures as well. Its development begins at pollination, and it may grow to a size enormously greater than the pistil or pistils of the flower. The wall of the ovary forms the *pericarp,* which encloses the seeds. The pericarp may be dry or fleshy, or both conditions may occur—in layers.

Fruits are classified as to type most simply by the following criteria: (a) whether they are from gynoecium alone or from gynoecium plus accessory structures; (b) whether they are from a single carpel or several —the latter either united or separated; and (c) whether they are derived from one flower or several flowers. This is a simple and logical classification, but a more detailed scheme is conventionally in use. The following outline includes most of its categories (see Fig. 8.17):

DRY FRUITS.
> Indehiscent Fruits (those in which the pericarp surrounds the seeds until they germinate).
>> *Achene.* Examples: dandelion, sunflower.
>> *Caryopsis* (*grain*). Examples: maize, wheat.
>> *Nut.* Examples: hazelnut, chestnut.
>> *Samara.* Examples: maple, ash.
>> *Schizocarp.* Examples: carrot, parsnip.
> Dehiscent Fruits (those in which the pericarp splits open along one or more sutures).
>> *Follicle.* Example: milkweed.
>> *Legume.* Examples: bean, pea.
>> *Capsule.* Examples: tulip, Jimson weed.

FLESHY FRUITS.
> *Drupe.* Examples: peach, cherry.
> *Pome.* Examples: apple, pear.
> *Berry.* Examples: gooseberry, orange, tomato.

AGGREGATE FRUITS (a collection of small drupelets together).
> Examples: blackberry, raspberry.

ACCESSORY FRUITS (in which the seeds are embedded in a modified receptable). Example: strawberry.

MULTIPLE OR COLLECTIVE FRUITS (derived from several flowers that occur close together). Examples: mulberry, pineapple.

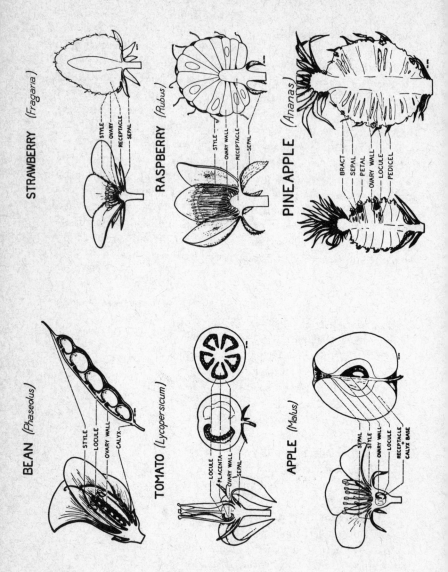

Fig. 8.17. Fruit morphogenesis. The development of various types of fruits from floral and accessory structures. See the outline in the text for the classification of the types represented. (Reprinted by permission from *Laboratory Directions for General Biology* by Gordon Alexander et al., published by Thomas Y. Crowell Company.)

*Condensed Descriptions of Representative Flowers and Fruits.* The following paragraphs summarize the characteristic features of the flowers and fruits of some of the most frequently studied plants. These illustrate various modifications in relations and numbers of flower parts. The first three types are of Dicotyledoneae, the last two of Monocotyledoneae. These examples are all of perfect flowers. Common plants bearing imperfect flowers are the maize (Indian corn), which is monoecious, and the willows and cottonwoods, which are dioecious. (The catkins of pussy willows are inflorescences.)

Buttercups have complete, regular, hypogynous flowers. The calyx has five sepals, and the corolla five petals. There is a large and indefinite number of stamens and pistils, spirally arranged. The fruit is an aggregate of achenes.

Beans and peas have complete, hypogynous flowers that are irregular (papilionaceous—"butterfly-shaped"). There are five sepals and five petals. The petals are quite varied; there is an upper one, the *standard,* two lateral ones, the *wings,* and two lower ones that together form the *keel.* There are ten stamens, one of these free but the others fused along the filaments. There is a single pistil, the ovary one-celled. The fruit is a legume.

The dandelion "flower" is actually an inflorescence of the type called a head. The individual flowers are epigynous. A double involucre, the outer scales reflexed, surrounds the base of the head. The calyx is a whorl of bristles, and the corolla of each individual flower is tubular at the base but spread out into a *ligule* (flaplike structure) above. All flowers of the dandelion head are ligulate. There are five stamens, united into a tube around the style. There is a single pistil, with a one-celled ovary. The stigma bears two lobes. The fruit is an achene borne by a parachutelike structure derived from the calyx.

The flowers of tulips are complete, regular, hypogynous. There are three petal-like sepals and three petals. The six stamens occur in two circles. The pistil is single, but it is composed of three carpels and is three-celled. The stigma bears three lobes. The fruit is a capsule.

In bluegrass, one to several flowers occur in a spikelet—which has two sterile *glumes* at its base. Each flower is enclosed by two bracts, a lower one, the *lemma,* and an upper, the *palet.* There are three stamens. Each flower has a single pistil, with a one-celled ovary and a two-lobed stigma. The fruit is a caryopsis or grain.

# Chapter IX

# THE PHYSIOLOGY OF MULTICELLULAR PLANTS

The nonreproductive functions of multicellular plants are related to several aspects of their activity. We here consider these functions under three general aspects: regulatory systems, metabolic functions, and growth. These are not mutually exclusive groups of functions, of course; they are closely interrelated. (Note: The details of cellular metabolism summarized in Chapter IV are not repeated in this chapter.)

## REGULATORY SYSTEMS

Though even one-celled organisms have some form of internal regulation, multicellular ones, which contain distinct organs and tissues, face increased regulatory problems. In these organisms regulation must involve communication between component parts—between cells, therefore. The nerve tissue that in animals provides rapid environmental analysis and communication within the organism is not present in plants. The mechanism in plants utilizes chemical messengers, *hormones,* which are also functional in animals. These provide slower patterns of communication, but it is these upon which plants are almost wholly dependent for internal regulation.

Hormones are organic chemical substances that are produced in certain cells and that influence the metabolic activity of other cells. Their production in plants is influenced by environmental factors. They are transported from the hormone-producing area to the target or influenced region primarily by diffusion, though some movement may occur in association with the translocation of dissolved foods. And because hormones are effective in extremely low concentrations such a system must involve either unstable hormones or enzymatic sequences that utilize or break down accumulating hormones; otherwise, of course, the concentrations of hormones would build up in the target region and their functions as regulators would be lost.

Some plant hormones are, by definition, vitamins—when we consider them in animal nutrition. What we think of as the B vitamins thiamine and niacin are, for example, hormones in plants, when they are manufactured in the leaves and exert their effect elsewhere. The terms hormone and vitamin do not refer to types of compounds; they are used only in a functional sense. Thus a substance may be a hormone under certain circumstances and a vitamin at other times. (In man the status of vitamin D is ambiguous because it may be manufactured in the skin but if not must be included in the diet.)

As previously stated, hormonal responses in plants are not rapid. This is because these responses are related primarily to growth. Plant movements due to hormones are essentially the result of differential cell elongation, which produces changes in the direction of growth. Other hormonal responses involve major morphological changes—the cyclic production of vegetative and reproductive organs at appropriate seasons. Not all plant movements are due to hormones, however; some of the turgor movements mentioned below may be induced by mechanical stimuli. All these regulatory activities are related to environmental stimuli, however, and they are fundamentally adaptive.

**Growth Regulators (Auxins) and Associated Tropisms.** Plant growing regions often produce hormones that diffuse to adjacent cells, where they control cell elongation and, in some cases, cell division and differentiation. One important group of such hormones includes the growth-regulatory substances collectively referred to as *auxins*. These are several natural and synthetic chemicals that produce similar activity, the major one being indole-3-acetic acid (abbreviated IAA). This is produced in the apex of a growing stem and moves down the stem in polar fashion, causing elongation of some cells. (It is also produced in root tips but in much smaller concentration.) It is required only in small quantities, and though the production is continuous the gradient of concentration decreases down the plant, due in part to an enzyme system that oxidizes IAA.

The production of auxin, and its effect, were learned from study of the oat coleoptile (the sheath around the young developing shoot). If the tip of the coleoptile is cut off and applied differentially to the cut surface (or if IAA is applied differentially to the cut surface) the cells under the point of application elongate more rapidly than the others. Thus, if the concentration of auxin is greater on one side of the stem than the other that side elongates more rapidly, producing a bending of the stem away from the region where auxin is more concentrated. Certain relatively continuous environmental factors—such as light and gravity—produce differential concentrations of auxins, and the resulting differential growth produces a turning of axial elements such as stems and roots.

These movements due to differential growth are called *tropisms*. They are *positive* if bending is toward the stimulus, *negative* if away from it. Various plant tissues respond differently, however, to the same auxin levels. The auxin level that induces elongation of stems, for example, inhibits elongation in roots.

Examples of tropisms include two of major importance, *phototropism* and *geotropism*. Phototropism is the result of differential growth with reference to a source of light. Light inhibits auxin production, and this results in lower concentrations in the lighted side and higher concentrations opposite—with bending of stems toward the light (positive phototropism) and of roots away from the light (negative phototropism). Geotropism is the growth response to gravity. The bending results from the concentration of auxins, by gravity, in the lower portions of an organ. Roots are positively geotropic, stems negatively so. Other tropisms in association with auxin and other plant hormones have been demonstrated: *chemotropism*, responses to chemicals; *hydrotropism*, response to water; *thermotropism*, response to heat; and *thigmotropism*, response to touch. Movements produced by differential growth but not with directional orientation (e.g., opening and closing of flowers) are *nastic movements*.

Auxins participate in the coordination of a variety of other plant activities. These include the dominance of apical buds, the formation of adventitious roots, the shedding of leaves and fruit through the formation of abscission layers, the maintenance of dormancy, and fruit development. Auxins have several commercial applications. They are used to insure the setting of fruit, the prevention of early fruit fall, production of roots in stem cutting, and control of potato sprouting. The differences in plant sensitivities to various synthetic growth regulators form the basis for the selective use of certain herbicides; 2,4-D is a synthetic hormone that, in certain concentrations, can be used to destroy unwanted broad-leafed plants in a grass lawn.

**Other Plant Growth Regulators.** In addition to auxins, at least two other groups of compounds function in plant growth regulation, the *gibberellins* and the *kinins*. Gibberellins are best known for their pronounced effects in stimulating cell elongation, and kinins for their effects on cell division and differentiation, but both have other effects as well. All three types of growth regulators are interrelated in functions. Their mechanism of operation is believed to involve the DNA-RNA-protein synthesis sequence (Chapter V). Another group of chemical regulators for which there is evidence are inhibitors, called *dormins* because of their apparent role in initiating and maintaining dormancy.

**Seasonal Regulation of Flowering and Vegetative Growth.** The relative length of light in daily cycles, the *photoperiod,* controls flowering and vegetative growth in certain types of plants. Individual plants respond differently to photoperiod, some flowering when exposed to short days and long nights (e.g., chrysanthemum) and others during long days and short nights (e.g., wheat). Temperature may also influence the time of flowering and the form of vegetative growth. The responses to photoperiod, periods of darkness, and temperature are variable, however, and some plants (e.g., dandelion) are relatively uninfluenced by these particular factors.

The photoperiod responses are controlled by a leaf pigment hormone, *phytochrome,* that is chemically changed by specific wavelengths of light. One form absorbs in the red region of the spectrum (660 m$\mu$) and is converted by red light into a far-red (730 m$\mu$) absorbing form, which is converted back to the other form by the far-red light. The pigment form and plant type control the response. Phytochrome controls a variety of plant responses such as seed germination, dormancy, stem and leaf growth, and flower production. (The hormone which responds to photoperiod, which is produced in leaves and causes the production of flowers, has been called *florigen.*)

**Movements Due to Turgor Changes.** Reversible changes, such as the closing of the insect-capturing leaves of the Venus's flytrap, the closing and drooping of leaves of the sensitive plant (*Mimosa*), or the opening and closing of stomata are due to changes in the turgor of particular cells. Some of these movements occur rather suddenly, as the drooping of leaves of the sensitive plant. Mechanical vibration initiates movement of water out of the cells that normally keep the bases of the leaf or leaflet rigid. This loss of turgor results in the sudden drooping of petiole or leaflet. The most significant plant movements caused by turgor changes are, however, those of the guard cells that regulate the size of the stomata. Their movements will be described in the next section, in connection with water relations.

# METABOLISM AND RELATED FUNCTIONS

The absorption and transport of water and inorganic nutrients, photosynthesis, biosynthesis, digestion, respiration, and excretion are considered in the following summary only in their special relations to multicellular plants. These functions in the individual cell have been considered in Chapter IV.

**Water Relations.** Plant metabolism requires free water. Water is one of the essential raw materials in photosynthesis, and it is the medium of transport of nutrients and foods in solution. In land plants water is absorbed through the roots, conducted upward in the stem, and evaporated through the stomata on the leaf surface. Undue loss of water is prevented by the presence of a cuticle over the leaves, a waxy, watertight layer that not only reduces evaporation but protects leaf cells from abrasion and, perhaps, from the damaging effects of ultraviolet light. This waxy envelope covers the epidermis except where the stomata are present.

ABSORPTION. Water is absorbed by the root hairs, first by imbibition of their cell walls. It is then conveyed through the root hair cells and the cortical cells of the root to the xylem vessels. A gradient of osmotic pressure, increasing in the root cells from outside in, is the most important factor in this movement.

WATER MOVEMENT. The water, which with its solutes constitutes the *sap,* is conducted upward in the xylem. It may rise to considerable heights—over 300 feet in certain trees. Several factors supplementing each other are involved in the rise of sap: (1) *root pressure*—the osmotic pressure that brings the water into the plant; (2) *capillary rise*—the tendency of a liquid that wets its container to rise to a height dependent upon the surface tension of the liquid; and (3) *cohesion* and *transpiration*—molecular attraction holding the column of liquid together, while its evaporation (transpiration) at the upper end tends to "pull" it upward. Of these factors capillary rise is least important; in a continuous column of liquid, cohesion and transpiration are undoubtedly the most important factors explaining the heights to which water rises in plants.

TRANSPIRATION (AND ACTIVITY OF STOMATA). Transpiration is the diffusion of water vapor from plants. (It is essentially controlled evaporation.) The rate of diffusion is influenced by plant structures (leaf size and shape; thickness and composition of the cuticle; size, number, and location of the stomata) ; by the opening and closing of stomata; and by environmental conditions (especially temperature and humidity differences between intercellular leaf spaces and the air surrounding the plant). The air spaces within the leaf are generally saturated with water vapor. Hence, as the air around the leaf heats in the sun, and its water vapor concentration becomes lower than that inside the leaf, water vapor diffuses out of the plant. Transpiration increases; and, if this continues too long, the water loss from the plant causes *wilting*—through a loss of cell turgor.

The mechanisms that prevent excessive loss of water are the *stomata* (also called stomates). These are openings in the epidermis, each opening between two *guard cells*. They are usually concentrated on the lower surfaces of leaves. The stomata function in gaseous exchange in general: they not only regulate water loss but are also the openings through which carbon dioxide—raw material in photosynthesis—enters the plant. Their activity is thus, appropriately enough, related to photosynthetic activity as well as to transpiration, and this is possible because the guard cells are the only epidermal cells containing chloroplasts.

The guard cells function in the opening and closing of the stomata through turgor changes. Their inner walls, next the opening between them, are thick, while the outer walls are thin. When turgor increases, the outer walls push out, pulling the inner walls apart; when turgor decreases, the outer walls retreat and the inner walls close the passageway. Control of the stomata is related both to water movement and to photosynthesis. When photosynthesis is not going on, as at night, carbon dioxide concentration increases in the cells as a result of respiration. The increased acidity (lowered pH) favors condensation of sugar to starch, resulting in a lowered osmotic or turgor pressure. Thus, the stomata tend to close at night, when photosynthesis is not going on. When the sun rises and photosynthesis begins again, the carbon dioxide is used up, the glucose concentration increases, and the turgor increases. This opens the stomata, permitting diffusion of carbon dioxide from outside into the intercellular spaces. At the same time, water relations are important. As the air warms in the sun and its water vapor deficit increases transpiration increases. If an abundant supply of water is available the guard cells remain turgid in sunlight, but when the loss of water reaches a rate greater than that at which it is conducted upward in the xylem it is drawn from the guard cells. As their turgidity falls the stomata close. Thus the shifts in turgidity of the guard cells regulate the inward movement of carbon dioxide into the plant and the outward movement of water vapor.

**Nutrients.** Nutrients are inorganic compounds essential to the life of the plant. (In animal physiology, the term is applied to the organic foods.) Nitrates, phosphates, sulfates, calcium, potassium, iron, and magnesium are important constituents of plant nutrients. Nutrients are absorbed in aqueous solution by the root hairs and are conveyed in the sap to the growing parts of the plant. They are used in the synthesis of foods and protoplasmic constituents and are involved in enzyme reactions.

Commercial fertilizers consist primarily of the nutrient ions most frequently deficient in the soil, or those needed for particular types of production. (Nitrogen is predominant in lawn fertilizers; it stimulates vegetative growth.) A "complete" fertilizer contains the three most important ions: nitrogen, potassium, and phosphorus.

**Photosynthesis.** As used by the biologist, the term photosynthesis is applied to the synthesis of carbohydrates through the action of chlorophyll in the presence of sunlight. It is, therefore, a function limited to green plants. (Chemically, however, any synthesis for which light is the source of energy is photosynthesis.)

SIGNIFICANCE. Photosynthesis in green plants is the most important agency for maintaining the fitness of the earth's atmosphere for the support of life. The oxygen used up in the process of respiration, in both plants and animals, is restored to the atmosphere through the process of photosynthesis.

THE PROCESS. Carbon dioxide is acquired by diffusion from the atmosphere through the stomata of the leaves. (Their action was described in the section on Water Relations.) Water, the other raw material for photosynthesis, is taken up by the root hairs from the soil and conveyed to the leaves in the xylem vessels. Through the catalytic action of chlorophyll, with sunlight as the source of energy, sugar is formed, and oxygen is given off as a by-product. Details of the process have been summarized in Chapter IV. The series of reactions involved may be summarized in the following shorthand equation:

$$6CO_2 + 12H_2O + \text{energy (from light)} \rightarrow C_6H_{12}O_6 + 6O_2 + 6H_2O$$

The simple carbohydrates formed in photosynthesis are the basic compounds converted into more complex carbohydrates, and into lipids, proteins, and nucleic acids.

CHLOROPHYLL. The green pigment involved in photosynthesis occurs in minute disc-shaped structures called *grana,* which comprise the chloroplasts. This pigment is of a compound nature, consisting, in higher plants, of four chemically different pigments: chlorophyll *a,* chlorophyll *b,* carotin; and xanthophyll. All contain carbon, hydrogen, and oxygen; the two chlorophylls contain nitrogen and magnesium in addition.

LIMITING FACTORS. The rate of photosynthesis is governed by that factor which is least available among those essential for the process. If either water, carbon dioxide, or light is available in below optimum quantity, the rate of photosynthesis will be limited by that factor. Thus, the amount of light limits the rate as the sun is rising or descending. In full sunlight, outdoors, carbon dioxide is usually the limiting factor. Temperature is important in governing the rate of the reaction also, as it has a significant influence on the rate of the "dark reaction" in photosynthesis.

**Biosynthesis.** The simple sugars manufactured in photosynthesis may be condensed to form double sugars and polysaccharides. They are also the basis for the synthesis of fats and other lipids. Some of their derivatives are combined with nitrogen-containing compounds to form amino acids and other organic compounds. Plants, unlike animals, are able to synthesize their own amino acids, using, as a source of nitrogen, nitrates (or nitrites or even ammonia) absorbed from the soil. Plants thus function as the ultimate source of amino acids for animal protein synthesis.

**Food Transport and Storage.** Manufactured food is conveyed from the leaves to other parts of the plant primarily in the phloem cells. (This movement is often called translocation, but that term is used with a different meaning in genetics.) Food is transported only in a readily soluble form, e.g., as a simple sugar. Starch may be formed and stored near to or distant from the place of photosynthesis, but, being insoluble, may not be transported. Its end products of digestion, being soluble, may be transported. All food material transported from one part of a plant to another must be in solution. It is carried primarily in the sieve tube cells of the phloem, but can sometimes be carried in the rising sap in the xylem. It can also be transported horizontally in vascular rays. The factors producing movement of dissolved food in the phloem are not clearly understood. Diffusion is one factor, of course, and cyclosis (cytoplasmic streaming) within the sieve-tube cells another.

**Digestion.** Digestion, which is hydrolysis of complex foods, occurs in the cells in any part of a plant in which food storage takes place. The material stored may be used where it is digested, or it may be transported to some other part of the plant. The end products of digestion are then assimilated or oxidized.

**Respiration.** Respiration in plants, as in organisms in general, is the oxidation of food for the release of energy. In the higher plants it is aerobic; that is, it requires free oxygen. The oxygen is absorbed through the roots (either from air spaces in the soil or in solution in water) and through the stomata of the leaves and the lenticels of the stem. The chief food used for oxidation is obtained through photosynthesis, the most commonly oxidized food being a hexose sugar like glucose. The steps in respiration are basically the same in all aerobic organisms, plants and animals. These have been described in some detail in Chapter IV.

**Excretion.** Plants have no specialized excretory systems as do animals. They do dispose of the waste products of metabolism, of course, but they do so by a variety of means. Some compounds, even organic substances, are known to be excreted by the roots, and an excess of carbon dioxide is released through various tissues, especially leaves; but many waste products are transported into particular cells where they are stored. These cells, for example those in the heartwood of a tree, die. They function primarily as recipients of excretory products that never leave the plant until the plant dies.

# GROWTH

Growth may be due to increase in cell size or increase in cell number, both types of growth being influenced by growth regulators previously mentioned. In plants, cell division takes place only in certain regions of undifferentiated cells, active metabolically, that are known as *meristematic tissue.* Meristems occur in the growing tips of roots, stems, and buds, and in the cambium. Of course, embryonic cells divide, too; meristematic cells are, functionally, merely persistent embryonic cells. Cells derived from meristem enlarge and differentiate. With only rare exceptions can these differentiated cells revert to meristem and divide again.

Major growth in all plants is from growing tips, and it is therefore in the main axis. This is called *primary growth. Secondary growth,* which results in the thickening of woody stems (the laying down of *annual rings* in the xylem and phloem) takes place by cell division in the cambium. Several types of regulators are involved in cell division, kinins and gibberellins in particular.

# Chapter X

# REPRODUCTION AND DEVELOPMENT IN MULTICELLULAR PLANTS

Reproduction in higher plants may take place by either *asexual* or *sexual* methods (Chap. V). In asexual reproduction only ordinary cell division, mitosis, occurs in the production of offspring, and only one parent is involved. (The cells of the parent may be either haploid or diploid, but no change in the chromosome complement occurs between parent and offspring.) In sexual reproduction, the process of meiosis invariably occurs between diploid generations, each of two diploid parents producing haploid gametes that unite in the process of fertilization. The progeny in sexual reproduction have a chromosome complement that differs from that of either parent qualitatively but not in number of chromosomes. There are two fundamental distinctions between asexual and sexual reproduction, in plants and in animals: (a) asexual reproduction is uniparental; sexual reproduction is biparental; and (b) cell divisions in asexual reproduction are all mitotic; in sexual reproduction, meiosis, followed by fertilization, must occur between generations.

## ASEXUAL REPRODUCTION

Reproduction of a unicellular organism by fission—ordinary cell division—is asexual reproduction. In multicellular organisms, the process may be almost as simple, but, in some cases, complex structures are involved. The following are forms of asexual reproduction in multicellular plants.

**Multiple Division.** Certain algae of definite cell number reproduce by a process of sudden cell division, each cell rapidly dividing until it forms as many cells as the original total. In this way an organism becomes as many separate organisms as it originally contained cells.

### Reproduction from Specialized Structures.

ASEXUAL SPORES. Single cells called spores may be produced in specialized structures, sporangia. If these spores are formed by mitosis and germinate into a new plant by mitosis, they are asexual spores. (It is necessary to distinguish between asexual spores and spores formed by meiosis. See paragraph headed Alternation of Generations.)

GEMMAE. Among the bryophytes occur special structures, the *gemmae,* which, after detachment from the parent thallus, are capable of growing into new thalli.

### Reproduction from Unspecialized Structures.

Artificial propagation of desirable horticultural varieties of plants is often carried out by root, stem, or leaf cuttings—more commonly stem cuttings. This technique is generally referred to as *vegetative propagation.* Strictly speaking, any form of asexual reproduction is vegetative reproduction.

ROOTS. The roots of some plants (e.g., wild plum, lilac) are capable of developing adventitious shoots from which new plants develop. Root cuttings are, in some species, more successful for artificial propagation than are stem cuttings.

STEMS. Artificial propagation by stem cuttings is familiar to all gardeners. Natural propagation of many species occurs—underground stems (*rhizomes*), surface runners (*stolons*), *bulbs,* and *tubers* being involved. Among plants commonly propagated by stem structures are iris (by rhizomes), strawberries (by stolons), tulips (by bulbs), and Irish potatoes (by tubers). In some species, drooping branches covered by soil develop new plants by the process of "layering." A new root system and vertical stem are developed from a node in these horizontal stems.

LEAVES. The ability of leaves to develop into complete plants is limited to a few species but does occur. African violets can be propagated artificially by this method, and, in several species of *Kalanchoe* ("maternity plants"), new plantlets develop asexually in the notches along the margins of the leaves.

## SEXUAL REPRODUCTION

Sexual reproduction is associated with the formation of specialized, haploid, reproductive cells and their subsequent fusion to form a fertilized egg. The mature reproductive cells that fuse are called *gametes.* If they are alike in appearance, that condition is known as *isogamy.*

Most plants are *heterogamous,* however, which means that the gametes are unlike in appearance. In such cases, the more active male gamete is called a *sperm cell;* the less active female gamete, an *egg cell* or *ovum.* The process of fusion is *fertilization,* and the cell formed as a product of fertilization is the fertilized egg or *zygote.* In plants exclusive of algae and fungi, the zygote develops, by repeated divisions, with differentiation, into an *embryo.* For this reason, the bryophytes and higher plants are collectively called the Embryophyta.

**Gamete Formation.** The haploid chromosome condition of gametes is, of course, acquired by meiosis (Chap. V). In plants, however, the haploid cells first formed in meiosis are not necessarily gametes. They usually divide by mitosis through many cell generations, producing a specialized plant body that may even be metabolically independent. Such a plant, whose cells are all haploid, eventually produces gametes, also by mitosis. This haploid organism in the life cycle of plants is absent from the life cycle of animals; hence, sexual reproduction in plants typically involves a much more complicated life cycle than occurs in animals. (See Alternation of Generations, below.)

**Fertilization.** The fusion of two gametes to form a zygote is fertilization. The diploid number of chromosomes is restored in this process. In higher plants, more than the fusion of two gamete nuclei is involved. Other nuclei may fuse to form cells involved in the nutrition of the embryo, the endosperm. Fertilization, with its related processes, is much more complicated in the higher plants than in any animals.

**Seed Development.** Two structures are important in seed formation. One is the *embryo* proper, which develops from the zygote; the other is the nutritive tissue, which, in angiosperms, is called *endosperm.* The mature seed contains a relatively large embryo, somewhat differentiated into organs; it may or may not contain endosperm. If endosperm is absent, the embryo has consumed it in the process of development.

**Germination.** The seed remains for a shorter or longer time in a *dormant* condition, following which, under proper environmental conditions, it develops into a young plant or *seedling.* The process of rupturing the seed coat and beginning growth is *germination.*

## ALTERNATION OF GENERATIONS

In the majority of plants the sequence of events in sexual reproduction is complicated by the occurrence of two different types of plants in the life cycle. (Fig. 10.1.) One type, the *gametophyte,* produces gametes. The gametes are haploid (*n*) in chromosome composition, but they are produced in the gametophyte by mitosis. All of the cells of the gametophyte are haploid, and the gametes are therefore produced by ordinary mitosis. After the gametes of opposite sex unite, in the process

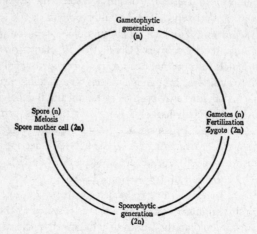

Fig. 10.1. A diagram illustrating the essential features of alternation of generations in plants. The haploid generation, the gametophyte, produces haploid gametes by mitosis. The gametes unite, by fertilization, to form the diploid zygote—from which the sporophyte generation develops. Haploid spores are produced by meiotic division of spore mother cells of the sporophyte, and a meiotic spore germinates into a gametophyte. (In this diagram, a single line indicates a single chromosome complement —the haploid condition; a double line indicates the diploid chromosome complement.)

of fertilization, they develop into a plant called a *sporophyte.* The cells of the sporophyte have the diploid (2*n*) chromosome number. This plant produces *spores,* but these are formed by meiosis and are, therefore, haploid. These spores, which we may call *meiotic spores,* germinate individually into gametophytes, and all of the cells of the gametophyte are, as previously stated, haploid. Thus the sexual cycle consists of an alternation between a diploid sporophyte that produces spores by meiosis and a haploid gametophyte that produces gametes by mitosis. The cycle appears to be a complex sexual cycle, but, because the spores develop individually, the sporophyte is sometimes called the asexual generation and the gametophyte the sexual generation. Such terminology will not be used in this book, however; the sporophyte and gametophyte will be considered two different phases in the complicated cycle of sexual reproduction. The expression asexual reproduction will be limited to reproduction not involving the sequence of meiosis and fertilization. (Note: The alternation of asexual and sexual reproduction in *Obelia,* as illustrated in Chapter XI, is not at all comparable with true alternation of generations; there is no alternation of haploid and diploid generations.)

## REPRODUCTION IN SPECIAL EXAMPLES

Characteristics of reproduction and development of the representative forms described in Chapter VIII are summarized here.

**Reproduction in the Fungi (Phylum Eumycophyta).**

BLACK BREAD MOLD (RHIZOPUS). Bread mold mycelia are of two different sexual types, but the two types may develop from two spores found in the same sporangium. When the hyphae of mycelia of different mating types come into contact they form zygotes. (Fig. 8.1.) Each zygote may contain many fusion nuclei rather than just one. From the zygote grows out a short hypha which forms a sporangium at its apex. The spores from the sporangium germinate into the two different sexual types of mycelia, in about equal numbers.

RED BREAD MOLD. Asexual reproduction occurs in *Neurospora* by the germination of asexual spores formed at the tips of hyphae, which are haploid. Sexual reproduction occurs when the tips of adjacent hyphae unite to form a zygote. The zygote divides by meiosis, and, with one additional mitotic division, forms eight haploid *ascospores* in an elongated sac, the *ascus.* Each ascospore can germinate into a new mycelium.

MUSHROOMS. The *basidia,* from which the spores develop, are cells containing fusion nuclei and, therefore, the diploid number of chromosomes. Each basidium nucleus divides by meiosis; the four nuclei go into four spores (*basidiospores*), each of which has, therefore, the haploid chromosome number. The mycelium and basidiocarp, coming from the spore, contain the haploid number of chromosomes.

### Reproduction in the Green Algae (Phylum Chlorophyta).

ULOTHRIX. The cells in a filament of *Ulothrix* are all haploid. They are potentially either sporangia, producing *zoospores* (each with four flagella), or *gametangia,* producing gametes (each with two flagella). Both zoospores and gametes are, of course, haploid. Although the gametes appear to be alike (are *isogamous*) they will not fuse unless derived from different filaments. The zygote divides by meiosis, form-ing zoospores, so there is no diploid stage except the zygote. Zoospores germinate into new filaments.

ULVA. In *Ulva* there is *alternation of generations* between a diploid sporophyte and a haploid gametophyte, these plants being similar in appearance. The sporophyte produces zoospores with four flagella each. The zoospores are formed by meiosis and are, therefore, haploid. Each zoospore germinates into a gametophyte, in which the cells are all haploid. Gametes of opposite sex may be identical (*isogamous*) or may differ in size (*heterogamous*), depending upon the species of *Ulva,* but the two gametes taking part in fertilization must come from differ-ent thalli. *Ulva* is *heterothallic,* in other words. The zygote divides mitotically, forming a diploid thallus; the latter divides by meiosis, thus producing zoospores and starting a new cycle.

OEDOGONIUM. The life cycle is similar to that of *Ulothrix* but some-what more advanced. Oedogonium is *heterogamous*. The large egg is fertilized in the cell (*oogonium*) in which it formed, by a male gamete from another specialized cell (*antheridium*). The zygote develops a thick wall. It germinates after a period of dormancy, forming, by mei-osis, four zoospores, each developing into a filament. Zoospores are also formed asexually from cells of a filament.

SPIROGYRA. *Spirogyra* filaments reproduce asexually. In addition, a form of sexual reproduction occurs. The contents of two similar cells of adjacent filaments may fuse in a fertilization process. (Fig. 10.2.) The fusion cell, a zygote, which is more or less resistant to drying, may develop later into a new filament. The cells of the filaments contain the haploid number of chromosomes; the zygote contains the diploid number. Its first divisions are meiotic.

### Reproduction in the Brown Algae (Phylum Phaeophyta).

ROCKWEED (FUCUS). The thallus of the rockweed is diploid. It does not form asexual spores. Sperm and egg cells develop in specialized cavities (*conceptacles*) at the tips of certain branches of the thallus; both escape into the water. There is no gametophyte stage, only haploid

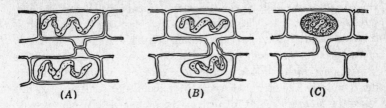

(A)    (B)    (C)

Fig. 10.2. Conjugation in *Spirogyra*. Three stages in the fusion of gametes. The gametes are the cell contents of cells of adjacent filaments. The walls separating the cells disintegrate, and one cell moves into the other through a conjugation tube. The zygote is shown in the upper cell at (C).

gametes, as in animals. Each sperm cell has two flagella. The fertilized egg divides into two cells: the holdfast and the thallus.

KELP (LAMINARIA). In the kelp the diploid thallus produces spores in sporangia, grouped in sori. The spores develop into small filamentous gametophytes that are dioecious. Fertilization occurs in the water, and a new sporophyte is formed. *Laminaria* has the type of alternation of generations in which the sporophyte is the conspicuous generation and the gametophyte is considerably reduced.

**Reproduction in Mosses and Liverworts (Phylum Bryophyta).** The gametophyte generation is dominant and vegetatively independent. The sporophyte is parasitic on the gametophyte. (Figs. 8.4 and 10.3.)

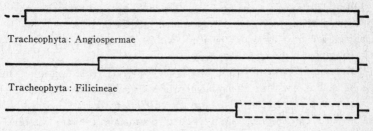

Tracheophyta : Angiospermae

Tracheophyta : Filicineae

Bryophyta

Fig. 10.3. Diagrams illustrating the evolution of the relations between gametophyte and sporophyte generations in plants. Single lines represent the haploid (gametophyte) generation, double lines represent the diploid (sporophyte) generation. The lengths of the lines correspond roughly to the relative prominence of the two stages in the life cycle. Generations represented by unbroken lines are independent; those represented by broken lines are parasitic on the other stage in the life cycle.

Moss.

*Sporophyte.* All cells of the sporophyte contain the diploid chromosome number because it develops from a fertilized egg by mitosis. It is parasitic on the gametophyte, remaining attached by its foot to the tip of the gametophyte stem. Spores develop in the capsule, from *spore mother cells,* each of which forms four spores by meiosis. Therefore, the spores are haploid. Each spore germinates into a filamentous network, the *protonema.* (At intervals in the protonema develop *gemmae,* or buds, each of which is capable of giving rise, asexually, to a gametophyte plant. When separated from each other, these plants are capable of independent survival.)

*Gametophyte.* All cells of the gametophyte contain the haploid chromosome number because it develops from a haploid spore by mitosis. The gamete-producing organs, *antheridia* and *archegonia,* occur at the apex of the same or different stems. The former are elongated, saclike organs in which are produced hundreds of two-flagellate sperm cells. The archegonia are bottle-shaped, each producing an egg cell in the expanded base, the *venter.* In rain or dew, the sperm cells swim to the archegonium, down the neck, and fertilize the egg cell. The sporophyte then grows from the zygote, which remains in the venter.

Liverworts. Reproduction in liverworts does not differ in principle from that in mosses. The gametophyte is broad and thalluslike, not erect; the antheridia and archegonia are borne on discs elevated above the thallus by stalks. The sporophytes may or may not be stalked.

**Reproduction in the Phylum Tracheophyta.** The gametophyte, which is small, may develop chlorophyll and be independent (in the Filicineae) or it may be parasitic within the sporophyte (in seed-bearing plants, the spermatophytes).

Reproduction in Club Mosses (Subphylum Lycopsida). In some club mosses (e.g., *Selaginella*) the spores are of two sizes, *microspores* and *megaspores.* The former develop in *microsporangia,* the latter in *megasporangia*—both types of sporangia in the same strobilus. Plants forming two kinds of spores are said to be *heterosporous.* Microspores in club mosses are formed in the same manner as spores in ferns; but, in the megasporangium, only one spore mother cell forms megaspores, the four megaspores formed growing to a large size by ingesting the nonfunctional spore mother cells. Microspores produce male gametophytes, which form sperm cells; megaspores produce female gametophytes, which produce eggs. The gametophyte is very small. Club mosses appear to bridge the evolutionary gap between ferns and seed plants.

REPRODUCTION IN HORSETAILS (SUBPHYLUM SPHENOPSIDA). Spores are borne in *sporangiophores,* which are arranged in whorls in *strobili.* In the most common species, the spores develop into either of two types of prothallia. One type produces antheridia, the other archegonia. The gametophyte generation is therefore *heterothallic,* the sperm cells and eggs being produced in different plants.

REPRODUCTION IN FERNS (CLASS FILICINEAE, SUBPHYLUM PTEROPSIDA). Both sporophyte and gametophyte are independent. (See Figs. 8.6 and 10.3.)

*Sporophyte.* The prominent, leafy fern plant is the sporophyte. Its cells are diploid. Spores are produced in sporangia in the *sori* from spore mother cells, each of which gives rise to four haploid spores. Upon germination, each spore forms a *prothallus.* Several independent sporophytes may develop from the nodes of a single rhizome. In some ferns, roots may develop from the tips of fronds in contact with the soil, and from them, other fronds.

*Gametophyte.* The cells of the prothallus contain the haploid chromosome number. The gametes are formed in antheridia and archegonia, both present on the underside of the same prothallus. The sperm cells are spirally twisted and flagellate; they swim to the egg cells in a film of moisture. The egg cell is fertilized in the archegonium. From the diploid zygote grows out the leafy sporophyte.

REPRODUCTION IN THE PINE (CLASS GYMNOSPERMAE, SUBPHYLUM PTEROPSIDA). The sporophyte is dominant and independent. The gametophyte is extremely small and is parasitic within the sporophyte. The specialized reproductive structures are *cones* (strobili). Fertilization is not dependent on water. *Seeds* are borne naked on cone scales.

*Sporophyte.* Staminate and ovulate cones (Fig. 8.7) are borne on the same plant. Staminate cones consist of scales called *microsporophylls,* each bearing two *microsporangia* or *pollen sacs.* The pollen sacs contain *microspore mother cells* (*pollen mother cells*). Each pollen mother cell divides by meiosis, forming four *pollen grains* or microspores. Each of these is, essentially, a male gametophyte. Ovulate cones, which are much larger than staminate ones, consist of scales called *megasporophylls,* each bearing two *ovules.* Each ovule contains one *megaspore mother cell,* surrounded by many spongy cells (the *nucellus*) and a coat (the *integument*). Through the latter extends an opening, the *micropyle.* The megaspore mother cell divides by meiosis to form a row of four cells, the basal one becoming the female gametophyte.

*Gametophyte.* Pollen grains (male gametophytes) are disseminated by wind. After reaching the ovulate cones, they are carried down the micropyle into a cavity (*pollen chamber*), where they germinate. In germination, each pollen grain forms a *pollen tube* containing two sperm nuclei. The pollen tube grows down to the archegonium. The female gametophyte grows at the expense of surrounding tissue. Within it develop several archegonia, each containing an egg nucleus.

Fertilization is accomplished by the fusion of one pollen-tube sperm nucleus with an egg nucleus. Several eggs in one gametophyte may be fertilized, but only one completes development. The nucleus of the zygote divides to form four nuclei. From each of these begins the development of an embryo. Usually only one embryo completes its development in each ovule, and, therefore, only one embryo occurs in each seed. The embryo consists of a *plumule* (shoot tip), *hypocotyl* (from the lower end of which the root develops), and numerous *cotyledons*. It is surrounded by haploid nutritive material and a hard seed coat. The seed is usually winged. In germination the seed coat is ruptured first by the developing root.

REPRODUCTION IN FLOWERING PLANTS (CLASS ANGIOSPERMAE, SUBPHYLUM PTEROPSIDA). To illustrate reproduction in the angiosperms only one account is given—a general one for the dicotyledons. (Figs. 8.14, 8.17, and 10.4.)

*Sporophyte.* The sporophytes bear micro- and megasporophylls in their flowers, the former being the stamens, the latter, the carpels. Both may occur in the same flower, in different flowers on the same plant (*monoecious*), or in flowers on different plants (*dioecious*). The anthers of the stamens contain cavities, the pollen sacs, in which pollen grains are formed. Pollen grains are haploid cells, each being one of four produced, by meiosis, from a pollen mother cell. As in gymnosperms, the cells produced by the pollen grain constitute, fundamentally, a male gametophyte. The carpels enclose cavities, the "cells," formed by the folding together of the megasporophylls or carpels. The seeds, therefore, are not borne naked on the megasporophylls as in gymnosperms, but are enclosed by them. The ovule grows from a point of attachment on the carpel which becomes the *placenta*. The ovule consists of a spore mother cell, nucellus, and integument with micropyle, as in gymnosperms. The spore mother cell divides meiotically, forming a row of four megaspores, the basal one of which persists, becoming the female gametophyte.

*Gametophyte.* (Fig. 10.4.) The pollen grains are carried to the stigma of the pistil by gravity, insects, wind, or other agency. There they germinate, each pollen grain forming a pollen tube containing two sperm nuclei and a tube nucleus. This constitutes the male gametophyte. The tube grows down through the style to and through the micropyle, carrying the sperm nuclei.

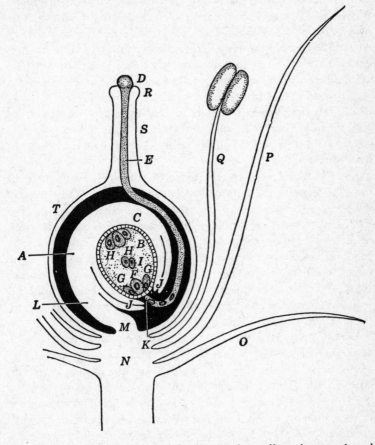

Fig. 10.4. Longitudinal section of flower, showing pollen-tube growth and ovule structure. (*A*) Ovule. (*B*) Embryo sac (megagametophyte). (*C*) Nucellus. (*D*) Pollen grain. (*E*) Pollen tube. (*F*) Egg nucleus. (*G*) Synergids. (*H*) Antipodals. (*I*) Polar nuclei. (*J*) Integuments. (*K*) Pollen tube entering micropyle, with one tube and two sperm nuclei. (*L*) Funiculus. (*M*) Placenta. (*N*) Receptacle. (*O*) Sepal. (*P*) Petal. (*Q*) Stamen. (*R*) Stigma. (*S*) Style. (*T*) Ovary. (Reprinted by permission from *General Botany* by Harry J. Fuller and Donald D. Ritchie, published by Barnes & Noble, Inc.)

Meanwhile, the female gametophyte or *embryo sac* has been formed. As the basal megaspore enlarges, consuming the three nonfunctional megaspores and surrounding cells of the nucellus, its nucleus divides, one daughter nucleus moving into each end of the cell. Each nucleus then divides twice, four nuclei being formed in each end. One nucleus from each end moves to the center, these two constituting the *polar nuclei*. One nucleus near the micropyle becomes the nucleus of the *egg cell*, while the other two become the nuclei of the *synergids*. The three nuclei at the other end form the *antipodal cells*. This eight-nucleated, seven-celled structure is the embryo sac, the mature female gametophyte.

*Fertilization.* One of the sperm nuclei unites with the egg nucleus, forming the zygote, which develops into the embryo. In most cases, the other sperm nucleus unites with the two polar nuclei to form a triploid *endosperm* nucleus. This is called *double fertilization*. The tube nucleus disintegrates.

*Seed Formation.* The embryo and endosperm develop at the same time, the former using part or all of the latter for food. The embryo of dicotyledons contains two cotyledons, that of monocotyledons, one (Fig. 8.9). In dicots, the axis of the embryo between the cotyledons consists of a lower region, the *hypocotyl,* which terminates in the embryonic root or *radicle,* and an upper region, the *epicotyl* (plumule) from which comes the young shoot. The single cotyledon of monocots is attached laterally to these axial structures. Seeds of the grass family, which are monocots, have membranes covering the tips of the epicotyl and radicle, the *coleoptile* and *coleorhiza,* respectively.

*Germination.* Under proper conditions of temperature and moisture, with sufficient oxygen, and after a period of dormancy, the embryo swells, bursts the seed coat, and develops into the mature sporophyte. The viability of most seeds is limited to a dormant period of a few years, but there are authenticated cases of seeds of lotus germinating after a dormancy of centuries. The stories of the germination of wheat from ancient tombs of Egypt are, however, myths.

# Chapter XI

# THE MORPHOLOGY OF MULTICELLULAR ANIMALS

The most primitive group of animals, the phylum Protozoa, has been treated in Chapter VI. All other phyla may be called Metazoa (or the phylum Porifera, sponges, may be separated from other multicellular animals as the Parazoa). The following phyla are of major importance, representatives being described in this chapter in some detail: Porifera, Coelenterata, Platyhelminthes, Nematoda (of Aschelminthes), Mollusca, Annelida, Arthropoda, Echinodermata, Chordata. In addition, the following phyla are mentioned and briefly characterized: Ctenophora, Nemertea, Rotifera and Nematomorpha (of Aschelminthes), Ectoprocta, Brachiopoda, Onychophora, Hemichordata. This list appears long, but several rather obscure and relatively unimportant phyla are not even mentioned; some authorities recognize over thirty phyla in the animal kingdom.

## THE PRINCIPAL CRITERIA FOR CLASSIFICATION OF ANIMALS INTO PHYLA

The phyla are distinguished from each other, not on the basis of a single set of characteristics, but by combinations of characteristics. These provide criteria for classification and at the same time are criteria of increasing complexity in an evolutionary series.

**Number of Cells.** The simplest phylum, the Protozoa, is distinguished from all others by the fact that each organism in it consists of a single cell. In general, too, members of the simplest phyla of Metazoa have fewer cells than do animals in more advanced phyla.

**Tissue Differentiation.** In metazoan phyla, the degree of differentiation of cells into tissue types varies. Porifera (sponges) have little tissue differentiation, but Coelenterata (jellyfishes and their relatives), though still primitive, show more tissue differentiation; and more advanced phyla have still more complex tissue variation.

**Kinds of Symmetry** (see below).

**Organ Differentiation.** Distinct organs are poorly developed in primitive metazoan phyla, but they are progressively more complex in more advanced phyla. They differ in kind as well as complexity, too, thus providing positive distinctions between phyla.

**Digestive System.** Intracellular digestion occurs in sponges, but in other primitive Metazoa there is a *gastrovascular cavity,* with a single opening serving as both mouth and anus. A complete digestive tract with separate mouth and anus is a more advanced type.

**Origin of the Mouth.** The digestive tract arises from an inner cavity produced by an infolding of cells in the early embryo. The mouth develops at or near the embryonic opening of this cavity (the blastopore) in animals called *protostomes,* while it develops as a new opening in those called *deuterostomes.* Biologists group the more complex animal phyla in two evolutionary lines based in part on this distinction.

**Number of Germ Layers.** Metazoa develop typically from either two or three germ layers (embryonic cell layers). Those of the former type are *diploblastic,* those of the latter type, *triploblastic.* The three embryonic cell layers are, in order from outside in, the *ectoderm, mesoderm,* and *endoderm.* The mesoderm is absent as a definite layer in diploblastic animals.

**Coelom or Body Cavity.** In the more advanced triploblastic animals a cavity develops within the mesoderm, and it is lined with mesoderm. This is the true *body cavity* or *coelom.* A false body cavity or *pseudocoel* characterizes certain primitive phyla of triploblastic animals. (Note: the word coelom in its various forms may also be spelled celom.) The most primitive triploblastic animals, as well as all diploblastic types, lack any kind of body cavity.

**Origin of the True Coelom.** Correlated with the difference in the origin of the mouth is a difference in coelom formation. Protostomes develop a coelom by a split in the mass of mesoderm cells, while in deuterostomes the coelom is formed by fusion of the cavities of several pouches from the forerunner of the digestive tract. The former type of coelom is called a *schizocoel;* the latter, an *enterocoel.*

**Metamerism.** Several phyla are distinguished by more or less complete segmentation of the body. This condition is known as *metamerism,* each segment being a *metamere.* The metameres show various degrees of independence from each other, but are subordinate for the most part to the whole organism.

**Differentiation of Organ Systems.** The presence, nature, and degree of development of particular organ systems vary from one phylum to another. Greater complexity is, in general, interpreted as meaning a more advanced or less primitive organism.

# DESCRIPTIVE TERMINOLOGY

**Kinds of Symmetry.** Animals may show no plane of symmetry, being asymmetrical (e.g., *Paramecium*), or they may possess spherical, radial, or bilateral symmetry to a greater or lesser degree.

SPHERICAL SYMMETRY. (The symmetry of a ball.) Any section through the center divides the organism into symmetrical halves. Examples: certain marine protozoa of the class Sarcodina.

RADIAL SYMMETRY. (The symmetry of a wheel.) Any vertical section through the center divides the organism into symmetrical halves. Examples: jellyfish, sand dollar.

BILATERAL SYMMETRY. (The symmetry of a plank.) Only one section, a vertical one in the longitudinal axis, divides the organism into symmetrical halves. Examples: earthworm, man.

**Planes and Sections of the Animal Body.** In a bilaterally symmetrical animal, a cut which follows any vertical plane parallel with the longitudinal axis is a *longitudinal section*. If exactly in the mid-longitudinal axis, it is a *sagittal section*. Sections in the vertical plane at right angles to the long axis are *transverse* or *cross sections*. Sections in the horizontal plane are *horizontal* or *frontal sections*. (Fig. 11.1.)

**Names of Directions.** The forward end is *anterior,* the opposite end, *posterior*. The back is *dorsal,* the lower surface, *ventral*. The sides are *lateral*. The point of attachment of a structure is its *proximal* end, the free end is *distal*. (Fig. 11.1.) All these terms are adjectives.

# MORPHOLOGY OF THE PHYLUM PORIFERA

**General Characteristics.** The Porifera are the sponges; they are mostly marine, but some occur in fresh water. They are sessile (attached) organisms. A sponge contains a central cavity (*spongocoel*) surrounded by a body wall penetrated by a series of canals of varying degrees of complexity. The internal cavity opens to the outside by the *osculum,* an opening at the distal end. A skeleton of *spicules* (of lime, glass, or spongin) supports the organism. A single "sponge" may be a colony of individuals all of which have budded from one parent. The embryonic

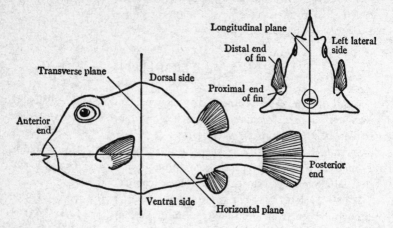

Fig. 11.1. Diagram illustrating descriptive terminology. An explanation of the names of *sections* occurs in the accompanying text. (The fish is a trunkfish, after Evermann and Marsh.)

germ layers are not comparable with those of other Metazoa. Reproduction takes place sexually and by budding.

**Example—*Grantia.***

General Characteristics. *Grantia* is a simple, cylindrical sponge, about an inch long, common along the Atlantic Coast. It has no specialized organ systems, except perhaps the unique canal system characteristic of sponges. (Fig. 11.2*A.*)

Canal System. This is the *sycon* type—i.e., *incurrent* and *radial canals* parallel each other and are connected by pores, the *prosopyles.* Incurrent canals have smooth lining cells; radial canals are lined with flagellated *collar cells* (or *choanocytes*), which ingest food. Water passes through incurrent canals, prosopyles, radial canals, central cavity (spongocoel), and out through the osculum. (Fig. 11.2*B, C.*)

Skeleton. Calcareous spicules of mono- and triaxial forms are present.

# MORPHOLOGY OF THE PHYLA COELENTERATA AND CTENOPHORA

Combined in one phylum by some authors, these are here separated but treated under a common heading.

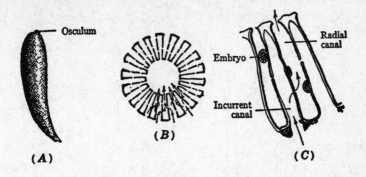

Fig. 11.2. A simple sponge (*Grantia*). (*A*) External appearance. (*B*) Diagram of a cross section. (*C*) Portion of body wall in cross section, enlarged. Arrows indicate the direction of water movement through the canals; water leaves the sponge through the osculum.

## Phylum Coelenterata (Cnidaria).

GENERAL CHARACTERISTICS. Hydra, jellyfishes, corals, and sea anemones are examples. Most coelenterates are marine; a few inhabit fresh water. They are radially symmetrical and diploblastic. The digestive cavity is a blind sac, the *gastrovascular cavity*, its opening functioning as both mouth and anus. This cavity is lined by a single cell layer, the gastrodermis. *Nematocysts* or stinging cells are present in the *epidermis*, the covering layer. Between epidermis and gastrodermis is the jellylike *mesogloea*. Among coelenterates, two body forms, the *polyp* and the *medusa*, may develop alternately during the life cycle.

*Polyp.* The general form is cylindrical, one end (proximal) attached, the distal end containing a mouth surrounded by *tentacles*. Colonial polyps form coral reefs. *Gonads* may be external or internal.

*Medusa.* The general form of a medusa is umbrella- or bell-shaped. The mouth is in the *manubrium*, which corresponds to the handle of the "umbrella." The digestive cavity has four main branches in the umbrella portion; these are the *radial canals*. They open into the *circular canal* around the rim. Tentacles hang from the margin of the umbrella surface. The gonads are suspended under the radial canals.

EXAMPLE—HYDRA. (Fig. 11.3.) *Hydra* is a freshwater coelenterate with the form of a polyp at all times. It is highly contractile and contains cells modified for conducting nerve impulses as well as for contraction. The gonads occur as swellings on the surface. Reproduction takes place sexually and by budding.

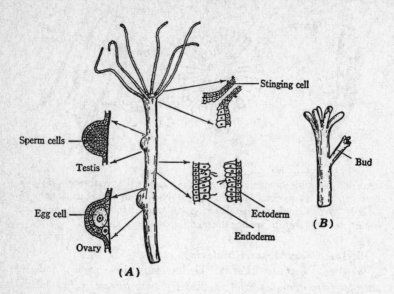

Fig. 11.3. *Hydra.* (*A*) Expanded individual, bearing gonads. Internal structures are shown in longitudinal section, enlarged, projected from four regions (indicated by arrows). (*B*) A partially contracted individual, bearing a bud.

EXAMPLE—OBELIA. (Fig. 11.4.) This is a marine animal whose small individuals occur together as a colony of polyps. The colony develops asexually from a zygote, and the polyps in the colony are of two kinds—*vegetative* (feeding) and *reproductive*. Each polyp is surrounded by a transparent covering, the *hydrotheca* about the vegetative polyps, the *gonotheca* about reproductive ones. All polyps are connected with each other by a hollow stem and branches. Reproductive polyps produce, by budding, medusae—which swim away when mature. They are sexually reproductive.

**Phylum Ctenophora.** All are marine. They are jellylike but not umbrella-shaped. Ctenophora are biradially symmetrical or walnut-shaped—hence are called "sea walnuts." They are also known as "comb jellies," from the characteristic eight rows of ciliated paddle plates, the plates of each row resembling the teeth of a comb. Mesoderm cells are present, the Ctenophora being the most primitive of triploblastic animals.

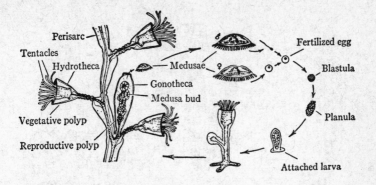

Fig. 11.4. *Obelia,* one of the colonial Hydrozoa. At the left, a portion of a colony, which reproduces asexually, by budding; at the right, diagrammatically shown, the sexual portion of the life cycle.

## MORPHOLOGY OF THE PHYLUM PLATYHELMINTHES

**General Characteristics.** This phylum comprises the flatworms—elongated, bilaterally symmetrical animals, flattened in the horizontal plane. Examples are *Dugesia* (a planarian), flukes, and tapeworms. Most flatworms are parasitic and some are pathogenic. They are triploblastic, but without a coelom. A gastrovascular cavity is present, except in tapeworms, which have no digestive system. Several organ systems are definitely differentiated.

**Examples.**

DUGESIA. (Fig. 11.5*A*.) *Dugesia* is a small, freshwater, free-living flatworm. Its body surface is ciliated. Eyespots are present near the anterior end. From the mouth, on the ventral side near the center, a protrusible pharynx (*proboscis*) leads into the digestive cavity, an intestine of three main branches, one anterior and two posterior. Each main branch has many small lateral branches. The excretory system consists of two longitudinal excretory tubes leading from numerous branches with *flame cells,* the whole system opening to the outside through dorsal *excretory pores.* There is a nervous system of two main longitudinal trunks, interconnected, and two anterior *ganglia* under the eyespots. The reproductive system is quite complex, the animal being hermaphroditic (male and female gonads in the same individual).

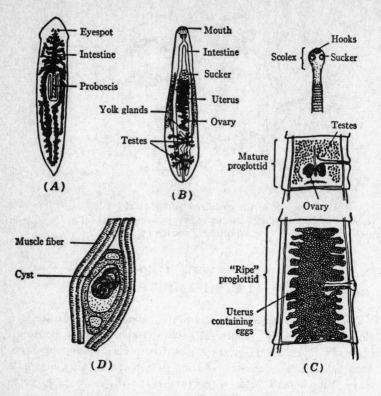

Fig. 11.5. Representatives of the phyla Platyhelminthes (*A-C*) and Nematoda (*D*). (*A*) *Dugesia*—digestive system in black. (*B*) *Clonorchis,* a liver fluke that parasitizes man in the Orient. (*C*) *Taenia,* a tapeworm that parasitizes man. The three drawings are all of the same scale of magnification. (*D*) *Trichinella,* encysted in muscle, cause of a serious disease in man and other animals.

LIVER FLUKES. (Fig. 11.5*B*.) Adult liver flukes are parasitic in the bile ducts of vertebrates. For attachment they have anterior and posterior suckers. There is a mouth at the anterior end, opening into the gastro-vascular cavity. Flukes are hermaphroditic, and they have complex life cycles. (See Chap. XIII, under Metagenesis.)

TAPEWORMS. (Fig. 11.5*C*.) Adult tapeworms are parasites of the alimentary canal of vertebrates. Each consists of a *scolex* (head)—with hooks for attachment, followed by a chain of several to many hundreds of *proglottids* formed by budding. Each proglottid is an hermaphroditic individual, having both ovary and testes. Excretory and nervous systems are reduced, and there is no digestive system. The larval forms, called *bladder-worms,* are acquired from a secondary host.

## MORPHOLOGY OF THE ASCHELMINTHES

Several groups of triploblastic, pseudocoelomate animals that are considered distinct phyla by some biologists are grouped into a single phylum, the Aschelminthes, by others. When considered a single phylum, the several groups become classes of the phylum Aschelminthes. We here treat these as distinct phyla, however. The most important of these groups, both in numbers of species and in economic significance, is the phylum Nematoda (which is called the class Nematoda when we recognize the phylum Aschelminthes). Others treated in this section as distinct phyla are the Rotifera and the Nematomorpha.

**Phylum Nematoda.** The Nematoda are the round- or threadworms. They have an elongated, cylindrical body, usually pointed at both ends, and with no segmentation. Some are important parasites—e.g., human hookworm and *Trichinella* (Fig. 11.5*D*)—but many are free-living. The digestive tract is complete, both mouth and anus being present. Nematodes have both an excretory and a nervous system. Reproduction takes place sexually, and nematodes are dioecious, i.e., the sexes are separate.

**Phylum Rotifera.** These are microscopic Metazoa, occurring chiefly in fresh water, where they are common. Circles of cilia at the sides of the mouth give them the common name *rotifers* ("wheel carriers"). They have a pseudocoel and a complete digestive tract. They are often observed in mixed laboratory cultures of protozoa and other freshwater organisms.

**Phylum Nematomorpha.** These are long, thin worms, without excretory or circulatory systems. The body cavity is a pseudocoel. They reproduce sexually and are dioecious. The "horsehair worm" is a member of this phylum. The adults are free-living, but the larvae are parasitic (usually in insects).

## MORPHOLOGY OF CERTAIN MINOR PHYLA

The relationships of the following phyla to each other and in an evolutionary sequence are not clear. All are triploblastic, but all are somewhat primitive in structure.

**Phylum Nemertea.** These are worms, mostly marine, related to the flatworms. They have a protrusible proboscis, which lies in an anterior sheath (considered by some the coelom) ; and they have a digestive tract with both mouth and anus and a blood-vascular system.

**Phylum Ectoprocta.** Animals of this phylum have the mouth enclosed in a crown of tentacles, the *lophophore,* a characteristic structure. A true coelom is present. Ectoprocts are small animals, usually colonial. They occur in freshwater and marine habitats but are much more abundant in the latter. (Mosslike in superficial appearance, these were formerly combined with a small, pseudocoelomate phylum, the *Entoprocta,* as the *Bryozoa.* The latter term is still used as a general term combining these unrelated forms.)

**Phylum Brachiopoda.** These are marine organisms, formerly more abundant than in present geological time. They live in bivalve shells, the shells occupying dorsal and ventral surfaces rather than lateral surfaces as in clams and mussels (phylum Mollusca). They possess a lophophore in common with the Ectoprocta and have a true coelom.

## PROTOSTOMIA (SCHIZOCOELA) AND DEUTEROSTOMIA (ENTEROCOELA)

As previously stated, biologists classify phyla in part on the basis of the origin of the mouth and the type of true body cavity or coelom. Phyla in which the mouth arises at or near the blastopore (protostomes) and in which the coelom is a schizocoel include Ectoprocta, Brachiopoda, Mollusca, Annelida, Onychophora, and Arthropoda. Phyla in which the mouth does not arise at or near the blastopore (deuterostomes) and in which the coelom is an enterocoel include Echinodermata, Hemichordata, and Chordata. (Each group includes additional minor phyla of primary interest only to the specialist.)

## MORPHOLOGY OF THE PHYLUM MOLLUSCA

**General Characteristics.** These are soft-bodied animals, usually protected by a calcareous shell of their own manufacture. They are marine, freshwater, and terrestrial. Common examples are: mussel, clam, snail, slug, nautilus, squid, octopus. Mollusks are triploblastic, nonmetameric, possess a coelom, have a complete digestive system, and have complex nervous, respiratory, circulatory, and reproductive systems. They are the most highly developed of nonmetameric animals. Recent evidence has led some biologists to believe they had metameric ancestors. Their pattern of embryonic development is indeed similar to that of the Annelida (which are metameric), but these two phyla probably diverged from a common ancestral stock before the establishment of metamerism. The Mollusca are more advanced in some characteristics than any group except the vertebrates, so it would be as logical to insert the account of this phylum after the Arthropoda as at this point. (Fig. 11.6.)

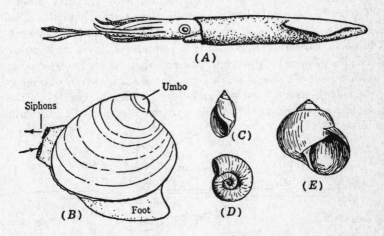

Fig. 11.6. Representatives of the phylum Mollusca. (*A*) Squid. (*B*) Hard-shelled clam or "quohog" (*Venus*). (*C*) Shell of *Physa*, a pond snail. (*D*) Shell of *Planorbis*, a pond snail. (*E*) Shell of *Helix*, the edible land snail.

**Example—Freshwater Mussel.**

EXTERNAL FEATURES AND GENERAL INTERNAL FORM. The mussel is enclosed in a *shell* consisting of two *valves*, on right and left sides, hinged at the dorsal side. Concentric lines of growth occur on each shell, centering at the *umbo*, the oldest part of the shell. Muscles, attached to the internal faces of the shells, close them and keep them together. The shells are lined wtih a delicate membrane, the *mantle*, the cavity within being the *mantle cavity*. A large muscular *foot* is capable of being extended between the valves at the anterior end. At its base is the large *visceral mass*, containing most of the organs, and suspended from this mass are four sheetlike, parallel *gill plates*. At the posterior end, the ventral *incurrent siphon* carries water into the mantle cavity. After the water passes over the gill plates, which filter out suspended food material, it is channeled out through the posterior dorsal *excurrent siphon*.

ORGAN SYSTEMS.

*Digestive System.* There is a mouth at the anterior end of the visceral mass, opening from the mantle cavity. The digestive tract consists of a short *esophagus, stomach,* long *intestine* coiled partly within the foot, and anus opening near the excurrent siphon.

*Respiratory, Circulatory, and Excretory Systems.* Oxygen, in solution, is taken up by the gills from water in the mantle cavity; carbon dioxide is given off. The gill filaments contain blood vessels. Blood from the gills passes to the *heart,* through one of the two *atria* and into the *ventricle.* The heart is enclosed in a *pericardium.* From the ventricle blood is pumped both anteriorly and posteriorly through two *aortae* to various parts of the body. It is collected in the *vena cava,* carried through the *kidneys,* thence to the gills and back to the heart. (Fig. 11.7.) The kidneys or nephridia drain through excretory pores into the dorsal part

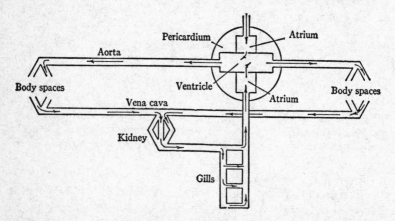

Fig. 11.7. Diagram illustrating the direction of blood flow in the freshwater mussel. The gill circulation is represented on only one side of the body.

of the mantle cavity, wastes being carried out through the dorsal siphon.

*Nervous and Sensory Systems.* Three pairs of ganglia (concentrations of nerve cells) are present, one pair near the esophagus, one in the foot, the other near the posterior end of the visceral mass. These are connected by longitudinal fibers, the two anterior ones also connected transversely. Sensory cells, probably sensitive to touch and light, occur along the margin of the mantle; an organ for detecting disturbance in equilibrium is present. Sense organs, however, are less well developed in mussels than in many other mollusks.

*Reproductive System.* Mussels are usually dioecious; they may be hermaphroditic. Their larvae are parasitic on the gills of fish.

**Land Snail.** The shell is spirally coiled. The snail, fundamentally showing bilateral symmetry, is considerably modified to conform to the coiled shell. A well-defined *head* is present, bearing two pairs of tentacles, the longer pair with *eyes* at their tips. A peculiar rasping organ, the *radula,* occurs inside the mouth. Land snails are air-breathing, respiration involving the wall of the mantle cavity, air entering through a small opening, the *pulmonary aperture.* They are hermaphroditic, but not self-fertilizing.

# MORPHOLOGY OF THE PHYLUM ANNELIDA

**General Characteristics.** These are the segmented worms. They are marine, freshwater, and terrestrial. Common examples are: earthworm, clamworm, leech. Annelids are triploblastic, metameric, have a coelom, and have complex digestive, nervous, excretory, circulatory, and reproductive systems—these in part or wholly metameric. The annelids are closely related, perhaps in an ancestral position, to the phylum Arthropoda.

**Example—Earthworm (*Lumbricus*).** (Figs. 11.8 and 11.9.) *Lumbricus* is the genus of some (but not all) common earthworms. It is the most widely studied example of the annelids but is not typical of the phylum in its complex reproductive system.

EXTERNAL FEATURES AND GENERAL INTERNAL FORM. An earthworm is elongate and cylindrical. Segmentation is visible externally as infoldings in the *cuticle.* Over one hundred metameres are present. The mouth is a slit at the anterior end, under a dorsal projection, the *prostomium.* The anus is at the posterior end. In sexually mature worms, a smooth swelling, the *clitellum,* occupies six or seven segments from about the thirty-second back. On each segment, except the first and last, there are four pairs of short bristles or *setae.* Internally, the coelom is divided into compartments by transverse partitions under external infoldings.

ORGAN SYSTEMS. (Figs. 11.8 and 11.9.)

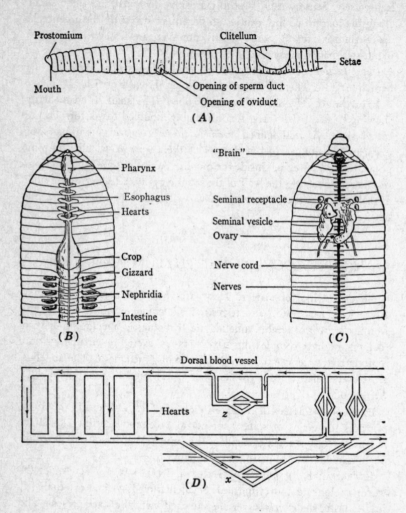

Fig. 11.8. The earthworm. (*A*) Side view of anterior third of body. (*B*) and (*C*) Dorsal views of dissections, with body wall spread out. (*B*) Digestive, circulatory, and excretory organs—the nephridia shown in only five segments, however. (*C*) Nervous and reproductive systems; the four testes lie under the spots marked *T*. (*D*) Diagram illustrating the directions of blood flow; the capillary systems marked *x, y,* and *z* are alternative routes; there is not general agreement concerning the directions of flow in the smaller vessels.

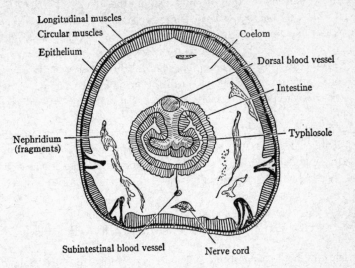

Longitudinal muscles
Circular muscles
Epithelium
Coelom
Dorsal blood vessel
Intestine
Typhlosole
Nephridium (fragments)
Subintestinal blood vessel
Nerve cord

Fig. 11.9. Cross section of a typical segment of the earthworm, diagrammatic.

*Digestive System.* The digestive system consists of: mouth; stout, swollen pharynx (segments 2–6); elongate, narrow esophagus (6–14); thin-walled *crop* (15, 16); muscular *gizzard* (17, 18); intestine (segments 19 to end of body); anus. The intestine has a dorsal internal fold, the *typhlosole,* running its length. (Fig. 11.9.) The esophagus has three pairs of *calciferous glands* along its sides. (Fig. 11.8*B.*)

*Respiratory and Circulatory Systems.* The earthworm breathes through the moist cuticle covering the entire surface, blood *capillaries* in the body wall taking up oxygen and giving off carbon dioxide. The blood-vascular system includes two main and three smaller longitudinal vessels. Blood flows forward in the dorsal vessel, through the five pairs of muscular hearts (segments 7–11) to the ventral vessel, thence to the body wall, thence to the dorsal vessel by way of the three smaller longitudinal vessels. There is disagreement concerning the direction of flow in the smaller vessels. (Fig. 11.8*D.*) The blood is red, containing hemoglobin in solution.

*Excretory System.* Paired *nephridia* are present in every segment except the first three and last one. Each nephridium drains material from the coelomic cavity anterior to it through its *nephrostome* and coiled *tubule* out through a ventral excretory pore. It also receives material by diffusion from blood capillaries surrounding the tubule. (Fig. 11.8*B*.)

*Nervous and Sensory Systems.* Sense organs in the skin are sensitive to contact and light. The nervous system is a ventral chain of ganglia, one in each segment beginning with the fourth, and an anterior *supra-pharyngeal ganglion,* the "brain," in segment three. Nerve cords around the pharynx connect the "brain" with the first ventral ganglion. Three pairs of *nerves* in each metamere arise from the ventral nerve cord (Fig. 11.8*C*.)

*Reproductive System.* Earthworms are hermaphroditic, but not self-fertilizing. (Fig. 11.8*C*.) Reproductive organs are complicated and variable in different kinds of earthworms. Some annelids are monoecious, the gonads duplicated segmentally.

## MORPHOLOGY OF THE PHYLUM ONYCHOPHORA

The Onychophora constitute a phylum of wormlike, terrestrial animals that are clearly intermediate between the Annelida and the Arthropoda. Its members have a soft integument and they have nephridia like those of segmented worms, but their respiratory system (tracheae) and circulatory system are like those of arthropods. They are metameric, but this condition is evident externally only in the paired, segmentally arranged legs. The genus *Peripatus* is a member of this phylum. *Peripatus* and the other members of the group occur in moist tropical and subtropical regions.

## MORPHOLOGY OF THE PHYLUM ARTHROPODA

The greatest variety in species and perhaps the greatest abundance in individual animals occurs in this phylum which numbers more known species than all other animals and all plants combined. Furthermore, the insects, man's greatest rivals economically, are the dominant representatives of the arthropods. For these reasons, the subdivisions of this phylum are treated here in some detail.

**General Characteristics.** Arthropods are segmented animals with jointed appendages and, in most cases, hard exoskeletons. They are found in marine and freshwater habitats and on land, most species being terrestrial. The following are common examples of the phylum: crayfish, crab, centipede, spider, mite, cricket, fly, beetle, butterfly, wasp. Arthropods are triploblastic and metameric. They have a true coelom but it is ill defined, and they have complex organ systems. They are, with the exception of the vertebrates, the most highly developed metameric animals—a peak in the evolution of segmentation with an external skeleton.

**Classes of the Phylum Arthropoda.** Five classes of arthropods are recognized in the following scheme. The Diplopoda and Chilopoda, here separated, are sometimes combined in one class, the Myriapoda. The phylum Onychophora is sometimes considered a class under the phylum Arthropoda, rather than a separate phylum as in this book. (Fig. 11.10.)

CLASS CRUSTACEA. Members of the class Crustacea are chiefly marine, but many live in fresh water and a few on land in damp places. They breathe through *gills.* Two pairs of *antennae* are present, and the larger crustacea have *compound eyes.* The appendages are derived from a common, *biramous* plan. Examples include: lobster, crab, pill bug, sow bug, water flea, barnacle.

CLASS DIPLOPODA. These are terrestrial forms with a cylindrical, elongated, wormlike body (but with a hard exoskeleton). They are distinctly segmented, with two pairs of similar legs on most segments. They have one pair of short antennae. The respiratory organs are *tracheae,* air tubes extending through the body. Millipedes are members of this class.

CLASS CHILOPODA. Members of this class are terrestrial forms with an elongated, flattened body, and with a hard exoskeleton. They are distinctly segmented, like the Diplopoda, but with only one pair of similar legs on each body segment. They have a single pair of long antennae. They breathe through tracheae. Centipedes constitute this class.

CLASS INSECTA. These are mostly terrestrial, though many species occur in fresh water and some are, rarely, marine. The body is divided into *head, thorax,* and *abdomen.* There is one pair of antennae—on the head; there are three pairs of legs attached to the thorax; and there are usually two pairs of *wings,* also attached to the thorax. The respiratory organs may be tracheae or, in aquatic forms, *tracheal gills.* The mouth parts are variously modified for chewing solid food or sucking plant and animal juices. Development may be fairly direct or may involve a complete transformation through several unlike stages, the former type being called *incomplete* and the latter *complete metamorphosis.* Members of this class are collectively called insects. Examples of insects are given in the following summary of the principal orders. (See Fig. 11.11.)

*Order Orthoptera.* Straight-winged insects with well-developed jump-

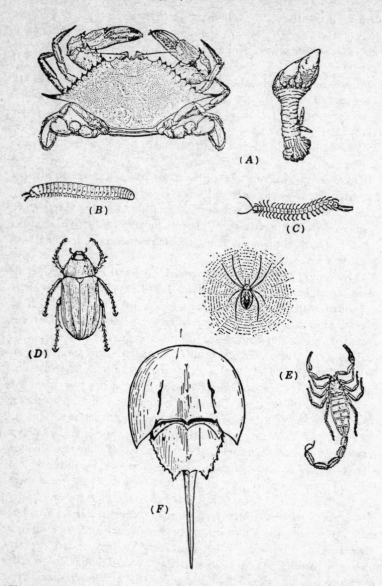

Fig. 11.10. Representatives of the Arthropoda. (*A*) Crustacea (crab and barnacle). (*B*) Diplopoda (millipede). (*C*) Chilopoda (centipede). (*D*) Insecta (beetle). (*E*) Arachnida (spider, scorpion). (*F*) Horseshoe crab. (*A-E,* copyright by General Biological Supply House, Chicago, and used by permission.)

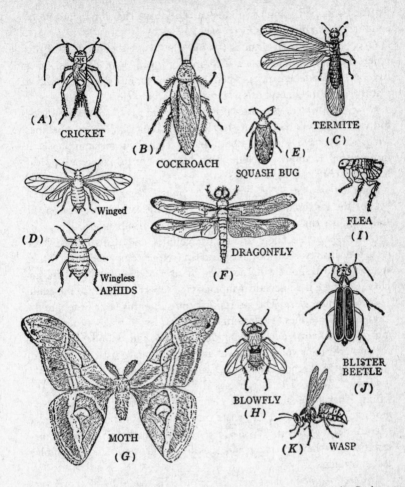

Fig. 11.11. Representatives of the principal orders of insects. (*A*) Orthoptera. (*B*) Blattariae. (*C*) Isoptera. (*D*) Homoptera. (*E*) Hemiptera. (*F*) Odonata. (*G*) Lepidoptera. (*H*) Diptera. (*I*) Siphonaptera. (*J*) Coleoptera. (*K*) Hymenoptera. (*B-H, J, K,* copyright by General Biological Supply House, Chicago, and used by permission.)

ing legs constitute this order. They have chewing mouth parts, and their metamorphosis is incomplete. Examples include crickets, katydids, and grasshoppers.

*Order Blattariae.* Running insects with flattened bodies, the cockroaches, form the order Blattariae. The four wings, when present, are flattened and overlap on the abdomen. They have chewing mouth parts, and their metamorphosis is incomplete.

*Order Isoptera.* The termites ("white ants") are the members of this order. They are small, soft-bodied insects that live in colonies. They have large, prominent jaws. Their four membranous wings are lost at maturity. The metamorphosis of termites is incomplete.

*Order Homoptera.* Small to medium-sized insects of somewhat varied form are members of this order. They have four, similar, membranous wings; they have sucking mouth parts; and their metamorphosis is incomplete. Examples include such diverse forms as cicadas, leaf hoppers, aphids, and mealy bugs.

*Order Hemiptera.* These are the true bugs. They are similar to members of the last group except that the forewings, which overlap on the abdomen, are thickened in the basal half—not membranous.

*Order Odonata.* These are the dragonflies and damsel flies, large predatory insects with movable head and large compound eyes. There are four membranous wings, similar in size. Metamorphosis is incomplete, but there is an aquatic nymphal stage superficially unlike the adult.

*Order Lepidoptera.* The insects that constitute this large order are the moths and butterflies. Their wings are covered with scales—these forming color patterns. Sucking mouth parts are coiled under the head. Metamorphosis is complete, involving egg, several larval (caterpillar) stages, a pupa (cocoon or chrysalis), and the adult.

*Order Diptera.* The Diptera, as their name suggests, have only two wings. They have sucking and piercing mouth parts, and their metamorphosis is complete. Flies, gnats, and mosquitoes belong to this order.

*Order Siphonaptera.* These are the fleas—small, wingless, jumping insects. They have piercing and sucking mouth parts and a complete metamorphosis.

*Order Coleoptera.* This very large order is made up of the beetles. They vary greatly in size but all have hardened forewings. The mouth parts are modified for chewing. Metamorphosis is complete.

*Order Hymenoptera.* The insects in this group have four membranous wings, the anterior pair being larger than the posterior pair. The head is free, and the mouth parts are chewing, but adapted for lapping. Metamorphosis is complete. Many colonial insects are in this order, which includes ants, bees, and wasps.

CLASS ARACHNIDA. The arthropods in this class have a body of two divisions, *cephalothorax* and *abdomen.* Many arachnids have a soft exoskeleton. The cephalothorax bears six pairs of appendages, four pairs being legs. Antennae, however, are absent. Arachnids have simple eyes but no compound eyes. They breathe through *tracheae, lung books,* or *gill books.* Most arachnids are terrestrial, but a few are aquatic. Ticks, mites, spiders, and scorpions are all members of the class Arachnida.

**Example of the Class Crustacea—Crayfish.** The following account applies equally well to *Cambarus,* the genus of crayfishes of the eastern United States, *Astacus,* the genus found in Europe and the Pacific Coast section of North America, and the lobster *Homarus.*

EXTERNAL FEATURES. (Figs. 11.12*A* and 11.13.)

*Divisions of the Body.* The major divisions are *cephalothorax* and *abdomen.* The former is covered by a hard shield, the *carapace,* which projects forward between the eyes as the *rostrum.* The first thirteen pairs of appendages, and the eyes, are attached to the cephalothorax; the remaining six are attached to the abdomen, which terminates in a horizontal fin, the *telson.* The abdomen is divided into segments, each protected dorsally and laterally by an arched skeleton. This consists of the dorsal *tergite* and two lateral *pleurae.* The ventral plates are known as *sternites.*

*Appendages.* Each appendage, except possibly the eye, is modified from the *biramous* type, which is ancestral and embryonic. (Fig. 11.12*C*.) It consists of the proximal *protopodite* and two distal branches, the *endopodite* (inner) and *exopodite* (outer). These are variously modified and reduced, as suggested in Table 11.1, but are essentially alike in origin and structure. They illustrate well what is meant by homologous structures. (Fig. 11.13.)

INTERNAL STRUCTURES. (Fig. 11.12*A, B.*)

*General Form.* A coelom is present, largely encroached upon by organs. There is a *hemocoel*—part of the blood system.

*Digestive System.* The digestive tract consists of a mouth, surrounded by several pairs of appendages—the "mouth parts," an esophagus, the stomach (with *cardiac* and *pyloric chambers*), intestine, and anus. The cardiac stomach contains grinding organs. Digestive glands (*hepatic glands*) pour enzymatic secretions into the pylorus of the stomach.

*Excretory System.* The two *green glands,* nephridial structures, open at the bases of the antennae.

*Respiratory System.* Feathery gills are attached to the basal segments of the second and third maxillipeds and the first four walking legs. Second and (in *Astacus*) third rows of gills are attached under the outer row. These are bathed by water in the *gill chamber* (the space under each side of the carapace) ; they contain blood vessels. The current of water through the gill chamber is maintained by the "bailer," a branch of the second maxilla.

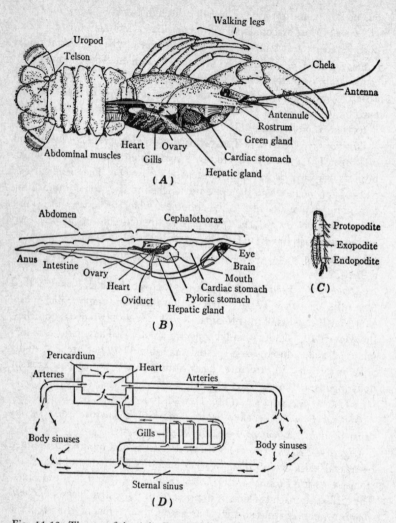

Fig. 11.12. The crayfish. (*A*) External features—part of the carapace removed to expose the internal organs in place. (*B*) Diagrammatic longitudinal section, showing internal organs in relative positions. (*C*) The fourth swimmeret, an appendage showing the primitive biramous condition. (*D*) Diagram illustrating the direction of blood flow.

*Circulatory System.* (Fig. 11.12*D*.) There is a dorsal heart, in a pericardium. Blood enters the heart through three pairs of *ostia,* valvular openings. It is pumped out through seven *arteries,* which empty into

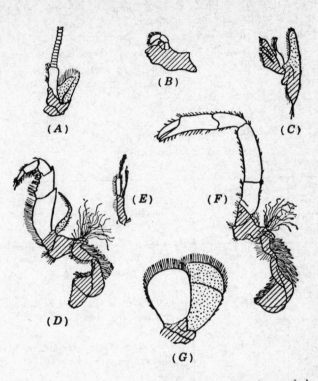

Fig. 11.13. Homology, as illustrated in the appendages of the crayfish, *Astacus*. Typical appendages from the left side of the body. (*A*) Antenna. (*B*) Mandible. (*C*) Second maxilla. (*D*) Third maxilliped. (*E*) Third abdominal appendage of female. (*F*) Second walking leg. (*G*) Uropod. The protopodite is shaded with parallel lines; the exopodite, with dots; the endopodite is unmarked. (Modified after Huxley.)

open spaces, the *sinuses*. These sinuses drain into the capillaries of the gills from which the blood enters the heart through the pericardium.

*Nervous System.* This consists of a dorsal brain, two *circumesophageal connectives,* and a ventral chain of ganglia. The first ventral ganglion is large, representing the fusion of several. Nerves branch from the brain and ventral nerve cord.

*Sensory System.* The sense of touch and the chemical senses of taste and smell are highly developed on the anterior appendages. Two *compound eyes* (consisting of many optical units, the *ommatidia*) are present, each eye on the end of a stalk. Organs of equilibration, *statocysts,* occur at the bases of the antennules.

*Reproductive System.* The sexes are separate (dioecious). The *testis* and *ovary* are bilobed. The testis empties by *sperm ducts* through pores at the base of the fifth pair of walking legs; *oviducts* convey eggs from the ovary to openings at the base of the third pair of walking legs. (Fig. 11.12*B*.)

### Table 11.1. Homologies of the Appendages of the Crayfish

| NUMBER OF APPENDAGE | NAME OF APPENDAGE | MODIFICATIONS FROM PRIMITIVE BIRAMOUS CONDITION (PROTOPODITE ALWAYS PRESENT) |
|---|---|---|
| 1 | Antennule | Ex- and endopodites both elongated sensory filaments. |
| 2 | Antenna | Endopodite a long sensory filament, exopodite a short basal blade. |
| 3 | Mandible | Exopodite absent; remainder a strong food-crushing organ with a palp. |
| 4 | 1st Maxilla | Exopodite absent. A thin organ lying just behind the mandible. |
| 5 | 2nd Maxilla | Exopodite constitutes "bailer" of gill chamber. |
| 6<br>7<br>8 | 1st Maxilliped<br>2nd Maxilliped<br>3rd Maxilliped | All parts present. Modified for manipulation of food. |
| 9 | Cheliped or 1st Walking Leg | Exopodite absent. Endopodite forms heavy pincher. |
| 10<br>11 | 2nd Walking Leg<br>3rd Walking Leg | Endopodite forms small pincher. |
| 12<br>13 | 4th Walking Leg<br>5th Walking Leg | No pincher on endopodite. |

Note: For appendages 10–13, a bracket to the right indicates "Exopodite absent."

### Table 11.1. Homologies of the Appendages
### of the Crayfish (continued)

| Number of Appendage | Name of Appendage | Modifications from Primitive Biramous Condition (Protopodite Always Present) |
|---|---|---|
| 14 | 1st Swimmeret | In female very small or absent; in male modified for transfer of sperm. |
| 15 | 2nd Swimmeret | In male modified for transfer of sperm; in female like next three appendages. |
| 16 | 3rd Swimmeret | All parts present—most nearly approach primitive form. Used in swimming and, in females, for egg attachment. |
| 17 | 4th Swimmeret | |
| 18 | 5th Swimmeret | |
| 19 | Uropod | All parts present, but broadened for swimming. Together with telson, constitutes tail fin. |

## Example of the Class Insecta—Grasshopper.

EXTERNAL FEATURES. (Fig. 11.14*A*.)

*Divisions of the Body.* Bodies of all insects have three major divisions: *head, thorax,* and *abdomen.* Each division consists embryonically of several segments. In the adult, three divisions of the thorax are visible: *prothorax, mesothorax,* and *metathorax.* In the grasshopper, the dorsal part of the prothorax is a saddle-like covering, the *pronotum.* The abdomen is divided into numerous segments, all distinct except the last few, which are fused and modified to form the *external genitalia.* About nine segments are clearly evident.

*Appendages.* The grasshopper has one pair of antennae, on the head, each antenna consisting of numerous segments.

The mouth parts are as follows: The *labrum* (upper lip) is suspended from the *clypeus.* Immediately under it are the two *mandibles*—heavy, chewing jaws. Under these are the two *maxillae,* each bearing a jointed *palp.* Between the basal parts of the maxillae is the pear-shaped *hypoglossus.* The divided lower lip, the *labium,* bears a pair of jointed palps. The maxillary and labial palps are sensory; the remaining parts function in food manipulation and mastication.

There are three pairs of legs, one pair on each of the three divisions

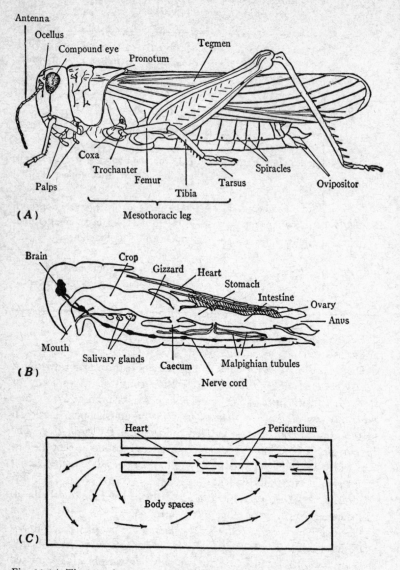

Fig. 11.14. The grasshopper. (*A*) External features, the segments of a typical insect leg being labelled in connection with the mesothoracic leg. (*B*) Diagrammatic longitudinal section, showing the organs in their relative positions. (*C*) Diagram illustrating the direction of blood flow.

of the thorax. The third pair (metathoracic legs) are strong and elongated, with a swollen femur, and are adapted for leaping. Each leg consists of the following segments, from proximal to distal end: *coxa, trochanter, femur, tibia, tarsus,* the latter having three segments in grasshoppers. The tarsus terminates in a pad (*pulvillus*) and two *claws.*

There are usually two pairs of wings, one each on meso- and metathorax, but they may be absent or reduced. The anterior pair, the *tegmina* (sing., *tegmen*), are hard, leathery, and more or less opaque. The posterior two are enlarged, membranous, transparent except when bearing color pattern or bands, and folded when at rest. There are numerous *veins* in the wings.

*External Genitalia.* In the male, the posterior tip of the abdomen is rounded below. Copulatory structures are largely concealed but include exposed lateral projections, the *cerci,* that are of varied form and serve as diagnostic characteristics of different species. In the female, the posterior end of the abdomen consists of the dorsal and ventral valves of the *ovipositor.*

INTERNAL STRUCTURES. (Fig. 11.14*B.*)

*General Form.* The coelom is reduced, and there is a haemocoel which is a part of the circulatory system.

*Digestive System.* The digestive system of a grasshopper consists of mouth, esophagus, crop, gizzard, stomach, intestine, and anus. These regions differ more in texture and thickness of walls than in internal diameter. Six double *caeca* (blind sacs) surround the gizzard and stomach and empty into the anterior end of the latter.

*Excretory System.* Numerous fine *Malpighian tubules* empty into the intestine at the point where the stomach joins it. These tubules, closed at their upper ends, drain excretory materials which diffuse from the haemocoel into the intestine.

*Respiratory System. Tracheal tubes* and *air sacs* are connected with the outside through lateral openings, the *spiracles.* Air is carried to cells as air, not as oxygen in solution. Carbon dioxide is partly excreted through other organs, however.

*Circulatory System.* There is a dorsal heart, blood entering it through five pairs of ostia from the pericardial sinus in which it is enclosed. The pericardial sinus receives blood from all parts of the body, from sinuses which blood reaches after being pumped out of the anterior end of the heart. (Fig. 11.14*C.*)

*Nervous System.* The nervous system consists of a dorsal brain, circumesophageal connectives, three thoracic and five abdominal ganglia. Branching nerves extend from the central system.

*Sensory System*. The antennae and palps probably contain organs of touch, taste, and smell. A *tympanic membrane* on the surface of the first abdominal segment is involved in detecting sound. (Sound producing organs of various types occur in grasshoppers.) Two compound eyes, made up of many ommatidia, are present, and there are three *ocelli*, or simple eyes.

*Reproductive System*. The sexes are separate. There are two testes or two ovaries present, in the dorsal part of the body cavity. Sperm ducts from the testes, or oviducts from the ovaries, open ventral to the anus.

### Example of the Class Insecta—Honey Bee.

KINDS OF INDIVIDUALS (CASTES). The males, which are fertile, are called *drones;* the fertile females are the *queens;* the sterile females are the *workers. Pollen baskets* are present only in workers. Drones have a broad abdomen; queens, a somewhat narrow, long abdomen.

EXTERNAL FEATURES.

*Divisions of Body*. These are the head, thorax, and abdomen. The thorax consists of prothorax, mesothorax, and metathorax. Six visible segments are present in the abdomen.

*Appendages*. The single pair of antennae consists of twelve segments in the male, thirteen in the female.

The following summary will describe the mouth parts. The labrum is broad and short, and the fleshy *epipharynx* projects beneath it. Two mandibles, horny jaws, lie lateral to the labrum and extend beyond it. Two maxillae extend much beyond the mandibles; they are covered with stiff hairs. Small maxillary palps are present. The labium is modified into a much elongated central portion, the *tongue,* and a large palp on each side.

There are, of course, three pairs of legs. Each leg consists of five joints—coxa, trochanter, femur, tibia, and a tarsus. The tarsus terminates in a pulvillus and lateral claws. The legs are covered with bristles and other structures used in collecting and carrying pollen and in cleaning pollen from the body. An antenna cleaner and an eye brush are present on the prothoracic legs. A spur on the mesothoracic legs is used to pry pollen from the pollen baskets, which are on the tibia of the metathoracic legs.

Two pairs of wings are present, both meso- and metathoracic wings being membranous. Veins divide each wing into "cells."

*External Genitalia*. The *sting,* present in queens and workers, is a modified ovipositor. In the male, copulatory apparatus is present.

## INTERNAL STRUCTURES. (Fig. 11.15.)

*General Form.* The form is typical of insects in general.

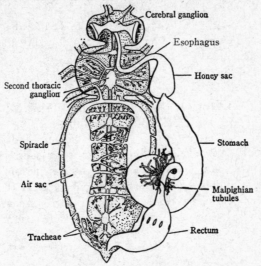

Fig. 11.15. Internal morphology of the honey bee, respiratory system partly eliminated on the right. (Redrawn from Henneguy, after Leuckart.)

*Digestive System.* This consists of mouth, esophagus, *honey sac,* stomach, intestine, *rectum,* and anus. The esophagus is elongated and thin; the honey sac is a globular region; the stomach is large and long; the intestine is narrow; and the rectum is enlarged. Digestive glands occur in the wall of the stomach.

*Excretory System.* Malpighian tubules drain into the intestine at its anterior end.

*Respiratory System.* Oxygen is not transported in the blood, but is carried in air in the tracheae. Lateral openings (spiracles) admit air into the tracheae and air sacs, branches of which convey it to all parts of the body.

*Circulatory System.* There is a dorsal heart in a pericardial sinus, receiving blood through five pairs of ostia. Blood is pumped anteriorly from the heart. After leaving the heart it diffuses through the hemocoel into the pericardial sinus.

*Nervous System.* A dorsal brain, a ventral chain of ganglia, and commissures or connectives around the esophagus constitute the major structures of the nervous system. Nerves branch from these.

*Sensory System.* Antennae probably bear end organs of smell, hearing, and touch. Taste organs are located near the mouth and on the tongue. There are two compound eyes and three ocelli.

*Reproductive System.* The sexes are separate. Fertilization is internal. There are two testes or two ovaries. The two ducts in either case join before reaching the genital opening.

## MORPHOLOGY OF THE PHYLUM ECHINODERMATA

Echinoderms are radially symmetrical marine organisms, possessing in most cases a spiny skin. Common examples are: starfish, brittle star, sea urchin, sand dollar, sea cucumber, crinoid. (Fig. 11.16.) They are triploblastic, possess a coelom, and usually have an anus. A *water-vascular system* constitutes a hydrostatic pressure system regulating

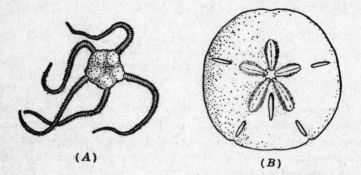

**(A)**    **(B)**

Fig. 11.16. Representatives of the Phylum Echinodermata. (*A*) Serpent star. (*B*) Keyhole sand dollar.

movements of the *tube feet,* locomotor organs characteristic of this phylum. The radial symmetry (usually on a plan of five antimeres) was probably derived from bilateral symmetry in the ancestral form. Similarity in stages of embryonic development suggests that echinoderms and chordates had a common ancestry.

## MORPHOLOGY OF THE PHYLUM HEMICHORDATA

The Hemichordata are elongated, burrowing, marine, wormlike animals that are closely related both structurally and embryologically to the

Fig. 11.17. (*A*) *Dolichoglossus*. (*B*) *Molgula,* a sea squirt. (Reprinted by permission from *General Zoology* by Gordon Alexander, published by Barnes & Noble, Inc.)

next phylum, the Chordata. They may have gill slits. Their nervous system includes both dorsal and ventral cords. An anterior projection from the digestive canal is sometimes considered a rudimentary notochord. *Dolichoglossus* (Fig. 11.17*A*) lives in burrows in sandy, intertidal areas along our Atlantic coast.

## MORPHOLOGY OF THE PHYLUM CHORDATA

Man and all his nearest relatives are members of this phylum, the majority of its members being vertebrates. The most highly developed of all animals belong to the phylum Chordata.

**General Characteristics.** Chordates are triploblastic, have a coelom, evident metamerism, and highly developed organ systems. An *endoskeleton* is always present at some stage. Its characteristic feature is the *notochord,* a gelatinous, longitudinal, dorsal rod, replaced in higher forms by a chain of vertebrae. Respiration always involves the *pharynx,* the region at the back of the mouth cavity, and *pharyngeal gill slits* are always present. There is a dorsal, hollow nerve cord.

**Major Divisions of the Phylum Chordata.** Three subphyla are recognized, the first two sometimes combined under the name Prochordata. The third subphylum, the Vertebrata, may be conveniently divided into seven classes, but more than seven are often recognized.

SUBPHYLA OF THE PHYLUM CHORDATA.

*Subphylum Urochorda (Tunicata).* Examples: sea squirt, sea pork. The notochord is present in the tail of the larva or tadpole; it is absent in the adult, which is sessile (attached). (Fig. 11.17*B*.)

*Subphylum Cephalochordata.* Example: *Amphioxus.* These have a fishlike, elongated body. The notochord is well developed, extending practically the full length of the body. (Fig. 11.18.)

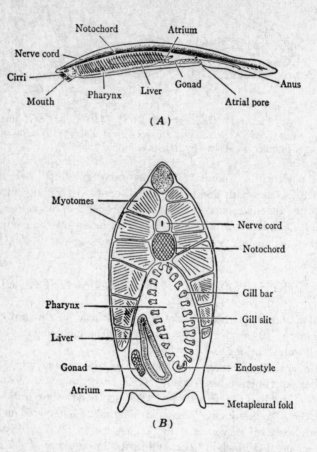

Fig. 11.18. *Amphioxus* (*Branchiostoma*). (*A*) Diagrammatic sketch of the right half, as exposed by sectioning in the sagittal plane (redrawn after Krause). (*B*) Diagrammatic sketch of a cross section through the posterior part of the pharynx.

*Subphylum Vertebrata.* Examples: lamprey, shark, perch, frog, turtle, bird, man. The notochord, if persistent, is surrounded by cartilage; if not persistent, it is replaced by a segmented, dorsal skeleton (*vertebral column*) of *vertebrae,* composed of cartilage or bone. Vertebrates have a central heart, red blood *corpuscles,* and an *hepatic portal system.* They usually have paired appendages with internal skeletal structures, but they never have more than two pairs of paired appendages.

CLASSES OF THE SUBPHYLUM VERTEBRATA.

*Class Agnatha.* Examples: lamprey, hagfish. These have a persistent notochord, a round mouth without jaws, and numerous *gill slits*. Paired fins are absent.

*Class Chondrichthyes.* Examples: shark, ray, sawfish. These are the cartilaginous fishes—the skeleton being of cartilage. Jaws are present (as in all succeeding classes). There are numerous gill slits. The body is covered with *placoid scales,* and paired *fins* are present. (Fig. 11.19.)

*Class Osteichthyes.* Examples: pike, perch, trout, cod, eel. In these fishes the skeleton is of bone or bone and cartilage. The gills are covered by a flap or *operculum,* leaving apparently only one gill slit. There is usually a covering of scales, but they are not placoid scales. Paired fins are usually present.

*Class Amphibia.* Examples: salamander, frog, toad, caecilian. (Fig. 11.21.) A smooth, moist skin, without scales, is characteristic of members of this group. Respiration is usually by gills in the young (which are aquatic), by *lungs* in the adult. There is a three-chambered heart (only two chambers being present in the heart in the three preceding classes). The eggs have a gelatinous covering.

*Class Reptilia.* Examples: turtle, snake, lizard, alligator. In reptiles, the skin is dry and covered with *epidermal scales*. Respiration takes place through lungs. There is a three-chambered heart (imperfectly four-chambered in some). The eggs have tough shells.

*Class Aves.* Examples: gull, snipe, pigeon, chicken, ostrich. Birds have a body covering of *feathers,* and the anterior appendages are modified for flight (with few exceptions). Birds are warm-blooded animals, having a four-chambered heart. Their eggs are covered with a hard shell.

*Class Mammalia.* Examples: opossum, rabbit, cat, whale, man. Mammals have a body covering of *hair*—which may be limited to certain areas or be almost entirely absent. *Mammary glands* (milk glands) are present in all female mammals. The young in most cases undergo embryonic development in the *uterus* of the female. This is *viviparous* development. (See Chaps. XIII and XVI.)

**Example of a Vertebrate—Dogfish Shark.** (In this and the following section, two examples of vertebrates are described, the dogfish shark and the frog. Repetition of similar details in the two accounts is avoided by cross references.)

EXTERNAL FEATURES. (Fig. 11.19.) The body is long, cylindrical, and tapering, the head somewhat flattened. The appendages are as follows: two median *dorsal fins,* a *caudal fin* (tail fin)—which is

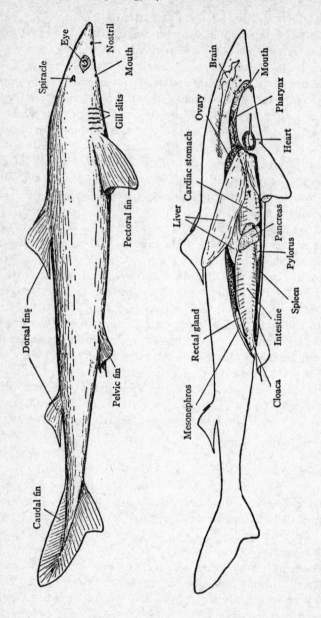

Fig. 11.19. Dogfish shark. Above, external features. Below, lateral view of a dissection in which the relative positions of organs are shown (one liver lobe has been moved to expose certain organs otherwise concealed by it).

heterocercal (the upper half being longer than the lower), a pair of anterior *pectoral fins,* and a pair of posterior *pelvic fins.* In the male, the pelvic fins are modified along the inner margins to form *claspers,* which are copulatory organs. The mouth is ventral in position. The nostrils are pits on the ventral margin of the head. The eyes are lateral in position. There are five gill slits on each side, posterior to and some-what ventral to the eyes. The *cloaca* opens between the pelvic fins. The body is covered with small placoid scales.

INTERNAL STRUCTURES. (Fig. 11.19.)

*General Form.* The coelom is divided into two compartments, a small anterior one enclosing the heart, the *pericardial cavity,* and a larger posterior chamber containing the rest of the viscera, the *peritoneal cavity.* The coelom opens to the outside through two *abdominal pores* at the posterior end.

*Skeletal System.* The brain and principal sense organs are enclosed or protected by a cartilaginous box, the *chondrocranium.* Beneath it is suspended the skeleton of the jaws and gill arches, this part of the skeleton being referred to collectively as the *visceral skeleton.* The verte-bral column is made up of a chain of cartilaginous vertebrae extending from the chondrocranium into the tail. Rudimentary *ribs* are attached to vertebrae. The fins are supported by *rays* of cartilage proximally and numerous fin rays of keratin distally. *Pectoral* and *pelvic girdles* sup-port the pectoral and pelvic fins, respectively, and connect them with the axial skeleton.

*Muscular System.* Regularly metameric muscle segments or *myotomes* occur through the body, but they are considerably modified in the head and appendages.

*Digestive System.* The mouth has jaws covered with *teeth,* these teeth being homologous with placoid scales. The pharynx opens into the respiratory chambers laterally, and the esophagus posteriorly. The esoph-agus merges into the cardiac portion, followed by the pyloric portion of the stomach, which bends anteriorly in a U. This empties into the *duodenum,* then into an expanded portion of the intestine containing a *spiral valve,* then into the *rectum* and the *cloaca,* the latter being a com-mon drainage channel for digestive, excretory, and reproductive systems. A *pancreas* and *liver* are present, their ducts (*pancreatic* and *bile*) emptying into the duodenum. A large *rectal gland* (function unknown, though of glandular structure) is present.

*Excretory System.* Two long narrow *mesonephroi* (sing., *mesonephros,* a type of primitive kidney) drain into the cloaca through the *Wolffian ducts.* These lie dorsal to the coelom, embedded in the muscles of the body wall.

*Respiratory System.* Gills are in lateral compartments of the pharynx, washed over by water from the mouth, which passes out through gill slits. Each half gill is a *demibranch.* Nine demibranches occur on each side, each containing blood capillaries.

*Circulatory System.* (Fig. 11.20.) The heart has two chambers, a dorsal *auricle* and a ventral *ventricle.* The blood is pumped forward from the ventricle through the *ventral aorta* to and through the gills (*afferent arteries* to gill capillaries to *efferent arteries*), to the *dorsal aorta,* and thence over the body through its branches. Blood enters the

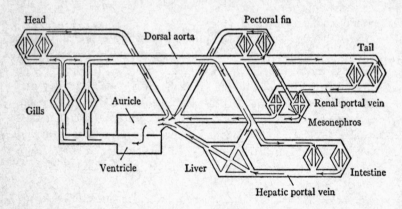

Fig. 11.20. Diagram illustrating the direction of blood flow in the dogfish shark. Applies equally well to all fishes.

auricle from the *sinus venosus,* draining the two anterior and two posterior *cardinal veins* and the two *hepatic sinuses* (from the liver). Blood reaches the liver directly, from the dorsal aorta, or indirectly, through the hepatic portal vein (which drains the capillaries of the intestine). The posterior cardinal veins drain the capillaries of the mesonephroi, which receive their blood from the posterior end of the body.

*Nervous System.* The nervous system consists of the brain and *spinal cord,* nerves as branches of each, and the *autonomic nervous system.* Nerve centers of the latter are more or less independent of the central nervous system. See Table 11.2 for names of embryonic brain divisions, common names of parts they include, and sources and distributions of the ten pairs of *cranial nerves.* Also see Figure 11.23*A. Spinal nerves* are distributed one pair to each body segment.

### Table 11.2. *Divisions of Brain and Their Relations to Cranial Nerves*

| DIVISIONS OF BRAIN | | CRANIAL NERVES | |
|---|---|---|---|
| *In Early Embryo* | *In Late Embryo or Adult* | *Number and Name* | *Peripheral Distribution and Function* |
| Prosen-cephalon | Telencephalon (Cerebrum) | I. Olfactory | Olfactory membrane; sensory. |
| | Diencephalon | II. Optic | Retina of eye; sensory. |
| Mesen-cephalon | Mesencephalon (Midbrain) | III. Oculomotor | Four eye muscles; motor. |
| | | IV. Trochlear | One eye muscle; motor. |
| Rhomben-cephalon | Metencephalon (Cerebellum) | | |
| | Myelen-cephalon (Medulla oblongata) | V. Trigeminal | Muscles and skin of face; mainly sensory. |
| | | VI. Abducens | One eye muscle; motor. |
| | | VII. Facial | Chiefly motor, to muscles of face. |
| | | VIII. Auditory | Cochlea and semicircular canals; sensory. |
| | | IX. Glosso-pharyngeal | Pharyngeal or gill region; sensory and motor. |
| | | X. Vagus | Heart, lungs, digestive tract; sensory and motor. |

*Sensory System.* The eyes differ from those of the frog (*q.v.*) chiefly in the absence of lids. The nostrils open to the outside, having no connection with the pharynx. They are lined with the *olfactory membrane,* from which nerves pass to the *olfactory lobes* of the brain. *Lateral line organs* probably function in hearing and in determining the direction of water currents. The *semicircular canals,* organs that function in detecting disturbances in equilibrium, are enclosed in the sides of the chondrocranium. There are three semicircular canals, an anterior vertical, a posterior vertical, and a horizontal one. Organs of taste are located in the skin.

*Reproductive System.* The sexes are separate and fertilization is internal. Embryos develop in the uterus of the female until able to swim. The female has two ovaries, in the anterior end of the abdominal cavity. Eggs, when mature, escape into the coelom, whence they are drawn into the oviducts through the *ostia* (funnel-like openings at the anterior ends of the oviducts). The lower end of each oviduct is swollen into a *uterus,* its walls extremely well supplied with blood vessels. The male has two testes, and mature sperm cells leave the body through the mesonephric or Wolffian ducts.

### Example of a Vertebrate—Frog.

EXTERNAL FEATURES. (Fig. 11.21.) The body of a frog is made up of the head, trunk, and two pairs of appendages, the hind pair much elongated. The skin is smooth and scaleless. The nostrils are anterior, the eyes dorsal, and the mouth a broad anterior slit. The *tympanic membrane,* flush with the skin, is behind the eye on top of the head. There are four fingers, plus a rudiment, on each hand, and five toes, webbed, on each foot. The *cloacal aperture* appears on the dorsal side between the hind legs.

INTERNAL STRUCTURES. (Fig. 11.21.)

*General Form.* The coelom of the frog has two compartments, the pericardial and the *pleuroperitoneal* (abdominal) cavities. The latter contains the lungs as well as the abdominal viscera.

*Skeletal System.* The *skull* of the frog consists of a small *cranium* (brain case) and a broad portion constituting the skeleton of jaws and face. The vertebral column consists of ten *vertebrae,* the first being the *atlas,* the ninth, the *sacrum,* and the tenth (greatly elongated), the *urostyle.* All vertebrae except the first, ninth, and tenth bear elongated *transverse processes,* the so-called "ribs." A *sternum* is present, connected with the rest of the axial skeleton through the pectoral girdle. The skull, vertebral column, and sternum constitute the *axial skeleton.*

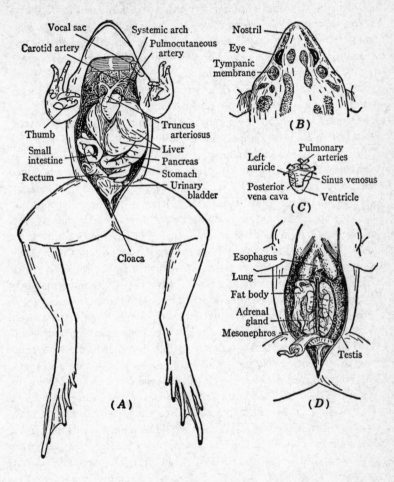

Fig. 11.21. The frog. (*A*) Ventral view, with abdominal wall removed to expose organs in place. (*B*) Dorsal surface of head. (*C*) Dorsal side of heart. (*D*) Ventral view, abdominal wall and some organs removed to expose organs of the dorsal part of the abdominal cavity.

The *appendicular skeleton* of the frog is made up of the skeletal structures of the two pairs of appendages and their attachments (girdles) to the axial skeleton. Each anterior or *pectoral girdle* consists of the following: *clavicle, coracoid, scapula, suprascapula* (cartilage). The posterior or *pelvic girdle* on each side consists of the *ilium, ischium, pubis*. The bones of each girdle are fused, and the right and left pelvic girdles are fused in the midline. Each pectoral appendage contains the following bones, in order from the proximal to the distal end: *humerus, radio-ulna, carpals, metacarpals, phalanges.* The corresponding bones in the pelvic appendage are: *femur, tibiofibula, tarsals, metatarsals, phalanges.*

*Muscular System.* Muscles show much greater differentiation from the primitive metameric pattern than in the shark, particularly in the appendicular muscles. No details are considered in this work.

*Digestive System.* In the mouth there are numerous fine teeth (*maxillary teeth*) along the upper jaw, as well as two groups of *vomerine teeth* in the roof of the mouth. The *tongue* is fleshy, bifurcate at the tip, and attached anteriorly. It is flipped forward in catching insects, which adhere to its sticky tip. The esophagus is a straight-walled, rather large tube, terminating in the stomach. The latter is an elongated muscular sac, curved toward the animal's left side. The intestine consists of a moderately long, coiled *small intestine,* the first portion of which, the *duodenum,* extends forward from the *pyloric* or posterior end of the stomach, and the *rectum,* a straight section merging into the *cloaca.* The latter is a common channel for digestive, excretory, and reproductive systems. The *liver* has three lobes, the median one containing a *gall bladder.* The pancreas lies in the *mesentery* (supporting tissue) between stomach and duodenum. Ducts from the pancreas and liver empty into the duodenum. In cross section, the intestine or stomach is seen to consist of four layers; these are, from the outside in, the *peritoneum* (continuous with the mesentery), the *muscular layer* (of smooth muscle, as in Fig. 7.1), the *submucosa,* and the *mucosa* (the lining).

*Excretory System.* Primitive kidneys or *mesonephroi* empty into the cloaca through the two *Wolffian ducts.* The kidneys lie dorsal to the peritoneum of the dorsal body wall. A *urinary bladder* is present on the ventral side of the cloaca; the Wolffian ducts do not empty directly into it.

*Respiratory System.* External and (later) internal gills are present in the tadpole. These are replaced by lungs at the time of metamorphosis, the adult frog breathing through lungs (and the skin). Each lung is a sac with numerous hollows in its walls. The lungs communicate with the outside through *bronchi,* a *larynx* (voice box), the pharynx, and the nasal passages between internal and external *nares.* The pharynx communicates with the larynx through the *glottis,* a longitudinal slit in the floor of the mouth behind the tongue.

*Circulatory System.* (Fig. 11.22.) The heart of the frog has three chambers, two *atria* (also called auricles) and one *ventricle.* Blood is

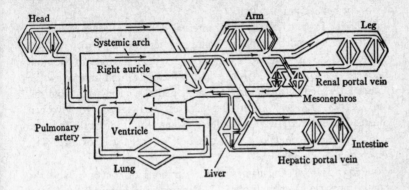

Fig. 11.22. Diagram illustrating the direction of blood flow in a frog. Applies equally well to toads, but not to all salamanders.

pumped out of the heart by the ventricle, leaving through the *conus arteriosus* (also called truncus arteriosus). The conus bifurcates anterior to the heart and then divides on each side into three trunks. From anterior to posterior branch these are the *common carotid artery,* the *systemic arch,* and the *pulmocutaneous artery.* Each common carotid divides immediately into an *internal* and an *external carotid artery,* both going forward into the head. Each pulmocutaneous artery sends branches to the lung and skin of the corresponding side. The systemic arches unite dorsally to form the *dorsal aorta,* after giving off branches to the arm and back. The principal arteries from the dorsal aorta are, in approximate order, the *coeliaco-mesenteric* (to stomach, liver, and intestine), *segmental* (to the body wall), *renal* (to the kidneys), and the two *iliacs* (to the legs).

Blood from the lungs returns in the *pulmonary veins,* emptying into the left auricle. All other blood enters the right auricle, passing through the *sinus venosus,* a large sac on the dorsal side. The sinus venosus receives two *anterior venae cavae,* with blood from the anterior part of the body, and one *posterior vena cava,* which originates between the mesonephroi and flows through the liver to the heart. Blood from the liver enters the posterior vena cava near the heart, through the *hepatic vein.* Blood reaches the liver from the *hepatic artery* (a branch of the coeliaco-mesenteric), or from the hepatic portal vein, draining the walls of the intestine. As in the dogfish, there is a *renal portal system,* carrying venous blood to the "kidneys" from the posterior part of the body. There is also a *ventral abdominal vein,* carrying blood from the legs to the hepatic portal system.

*Endocrine System.* The glands of internal secretion produce *hormones.* The *pituitary gland,* beneath the brain, secretes hormones that regulate growth and metamorphosis, sexual cycles, pigment changes in the skin, and water relations. The *thyroid gland,* which is behind and below the tongue, is related to metamorphosis and, perhaps, general metabolism. The *islets of Langerhans,* in the pancreas, are related to glycogen storage. The *adrenal glands,* embedded in the ventral side of the kidneys, have few known functions in frogs but are essential for life. The gonads secrete sex hormones.

*Nervous System.* The names of the divisions of the brain and the spinal nerves are given in Table 11.2. Differences in proportions of parts between brains of dogfish, frog, and man are indicated in Figure 11.23. Spinal nerves of the frog are ten in number; the first three have interlacing fibers constituting the *brachial plexus;* the seventh, eighth, and ninth form in the same way the *sciatic plexus.* Each plexus originates in an enlarged portion of the cord, the *brachial* and *lumbar enlargements.*

*Sensory System.* The *eyes* have upper and lower *eyelids* and a transparent third eyelid, the *nictitating membrane.* The *eyeball* is approximately spherical; its outer face is covered with the thin transparent *conjunctiva* (reflexed under the lids) and under it the transparent but thicker *cornea.* The cornea is continuous with the *sclera,* the outer opaque covering of the sides and back of the eyeball. Under the sclera lies the *choroid,* which, in front, merges into a doughnut-shaped shelf not in contact with the cornea, the *iris.* The hole in the iris is the *pupil.* The *crystalline lens* fits against the back of the iris, closing the pupil. The back of the inside of the eyeball is lined with nerve tissue, the *retina,* directly continuous with the *optic nerve.* The cavity in front of the lens and iris contains the *aqueous humor;* that behind, the *vitreous humor.* The eye is moved by six muscles. (See Fig. 15.6.)

The *ear* includes organs both for hearing and the sense of equilibrium, the latter in the semicircular canals (see account of dogfish). A *tympanic membrane* (with no external ear) conveys impulses set up by sound waves to the *columella,* a thin bone in the *middle ear,* which transmits impulses to the *cochlea,* the organ of hearing. The middle ear communicates with the pharynx through the *Eustachian tube.*

*Reproductive System.* The sexes are separate and fertilization is external. The female has two ovaries, ventral to the "kidneys." Mature ova escape into the coelom and are drawn into two oviducts, which lead

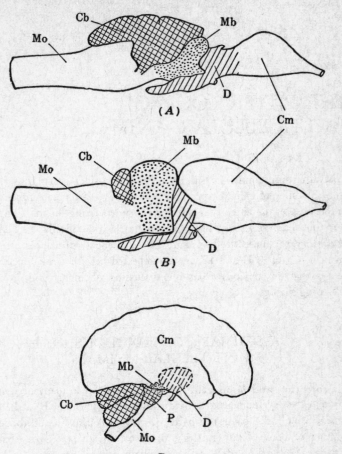

Fig. 11.23. Diagrammatic side views of the brains of (A) the dogfish shark, (B) the frog, and (C) man. From anterior to posterior (right to left) the five major divisions distinguished are: telencephalon, diencephalon, mesencephalon, metencephalon, myelencephalon. Abbreviations: *Cb,* cerebellum; *Cm,* cerebrum; *D,* diencephalon; *Mb,* midbrain; *Mo,* medulla oblongata; *P,* pons.

to the cloaca. Gelatinous envelopes are deposited around eggs in the oviducts. In the male there are two testes, from which the sperm cells reach the cloaca in the Wolffian ducts.

# Chapter XII

# THE PHYSIOLOGY OF
# MULTICELLULAR ANIMALS

In discussing animal physiology one may consider either the physiology of different animal types or the physiology of separate functions. In this chapter the latter plan is followed, summarizing the most important functional relations. The material bearing specifically on human physiology is summarized in Part Three. General principles already given in Chapters II and IV are not repeated here, nor are functional details that are obvious from the discussion of morphology in the preceding chapter.

## A SUMMARY OF FUNCTIONS IN
## MULTICELLULAR ANIMALS

**Protection and Support.** Protection is attained in many organisms through the development of a hard exoskeleton. This may be a shell of lime (in mollusks, corals), a hard covering of chitin (in anthropods), a series of bony plates (turtles), or a coat of scales (fishes, snakes, lizards). Feathers (birds) and hair (mammals) are protective, as well as means of conservation of heat for warm-blooded animals. Protection of internal organs may also be given by an endoskeleton; e.g., in vertebrates the cranium protects the brain, and the ribs protect the thoracic viscera. Both endo- and exoskeletons help maintain body form, furnish support for organs, and provide places for the origin and insertion of locomotor muscles.

**Movement and Locomotion.** Locomotion in Metazoa is usually due to contraction of muscles, but it may be accomplished in some cases by cilia (e.g., in *Dugesia*). Muscle movements are basic to such diverse functions as body movement, posture, and expression; heat regulation through shivering; elevation or depression of feathers or hair; digestive activity through peristalsis and the regulation of sphincters; accessory respiratory movements producing air movement into and out of lungs and water movement over gills; blood flow, through heart action and the regulation of the diameter of blood vessels; expulsion of secretory products from glands and reservoirs; regulation of the activity of sense organs (e.g., eye movement and focus); and control of the birth process in viviparous animals.

MECHANICS OF MUSCLE CONTRACTION. Muscle fibers are typically elongated. They contain long, thin organelles called *myofibrils,* which are accompanied by numerous mitochondria. The myofibrils are composed of protein strands (*myofilaments*), which include the protein groups *actin* and *myosin.* During contraction, actin and myosin combine, and this results in a shortening of the myofibrils. Two theories have been offered to explain the mechanism of the shortening. Parallel filaments may slide past each other, or there may be internal molecular folding. Whatever the process of internal change, the shortening is transmitted mechanically to connective tissue and tendons. (Muscle contraction involves no change in total volume of tissue; the muscle thickens at the same time it shortens.) If the shortening takes place across a skeletal joint this results in movement at the joint, the movable part acting as a mechanical lever. (See Fig. 14.2.)

CHEMISTRY OF MUSCLE CONTRACTION. Muscle cells convert chemical energy into mechanical energy. This conversion is thought to be triggered by depolarizing events in the muscle cell membrane. In the case of vertebrate skeletal muscle this depolarization is initiated by a nerve impulse (see below). When a source of "trigger energy" is released, actin and myosin combine to form *actomyosin,* this resulting in a shortening of the myofibrils. The process requires ATP and calcium ions, and chemical energy transformed into work comes through the conversion of ATP to ADP. Other muscle compounds, *phosphocreatine* (in vertebrates) and *phosphoarginine* (mainly in invertebrates), act as energy storage units. (These are collectively called *phosphagens.*) These compounds replenish the ATP store by transferring a phosphate group to ADP. Glycogen is the muscle carbohydrate from which the energy for ATP formation is derived (see p. 44). Under anaerobic conditions, lactic acid is formed, this process generating a low-level supply of ATP. By means of ATP, phosphagens, and anaerobic metabolism, the muscle can contract for some time in the absence of oxygen. However, the recovery phase that "retriggers" the muscle, reseparating actin and myosin, involves oxygen and results in the redevelopment of the ATP store and the phosphate storage units. In the recovery, most of the lactic acid is reconverted into glycogen, the energy for this process coming from the complete oxidation of part of the lactic acid. A muscle becomes fatigued and ceases to contract if contractions occur so frequently that oxidative recovery cannot take place.

**Digestion.** Organic material entering the digestive system of higher Metazoa is (1) broken down by mechanical actions, (2) enzymatically digested into small diffusable compounds, and (3) transferred through the membranes of the digestive tract into the body fluids.

MECHANICAL PROCESSES INVOLVED IN DIGESTION. These include *mastication,* which may occur in the mouth (grasshopper, crayfish, man) or farther along in the alimentary canal (in gizzard of earthworm, stomach of crayfish, or gizzard of bird). There may be no mastication (frog). *Swallowing,* the next process, is voluntary; but food movements through the alimentary canal, largely by *peristalsis,* rhythmic contractions, are involuntary. Food is retained in the stomach of vertebrates until thoroughly acidified, a small bit at a time being then passed through the pyloric valve. *Defecation* (egestion), discharge of undigested or unabsorbed materials, is voluntary.

CHEMICAL PROCESSES INVOLVED IN DIGESTION. Digestive enzymes catalyze the hydrolysis or breakdown of large organic molecules. These enzymes are referred to as *carbohydrases* if they digest carbohydrates (breaking down polysaccharides or disaccharides to simple sugars); *proteases* if they digest proteins (breaking down proteins and polypeptides to peptides and amino acids); and *lipases* (*esterases*) if they digest fats (breaking them down to glycerol and fatty acids). The digestive enzymes, as well as lubricants, mucus, emulsifying agents, and acid, are variously produced by gland cells of the digestive system. Glands that open into the oral cavity are called *salivary glands,* and their secretion is called *saliva.* (They may produce toxins, as in poisonous reptiles.) The other digestive glands and gland cells of the lining of the gut produce *digestive juice,* modified variously in different animals and in different compartments of the digestive system. Relatively complete knowledge of digestive action is lacking for many invertebrates. The description of the digestive enzymes in man (Chap. XV) will serve as a pattern for digestive action.

FOOD ABSORPTION. The digested food passes through the epithelial lining of the digestive tract by diffusion or, in some cases, by phagocytosis. In the higher phyla, the circulatory system is involved in food absorption. The part of the gut where absorption occurs often shows an increase in surface area—through elongation, the presence of branches with blind endings (*caeca* or *diverticuli*), or the presence of fingerlike protrusions (*villi*). In many mollusks and arthropods digestion and absorption take place largely in the cavity of the branching midgut glands (*hepatopancreas*), and the gut itself is relatively short.

**Respiration.** Respiration, in its broadest meaning, includes all processes bringing oxygen into the body and removing carbon dioxide, together with the process of oxidation within the cells. (In common usage the word respiration is limited to breathing, but here we recognize breathing as only a part of the process.) Cellular respiration has been described in Chapter IV. Practically all animals require free oxygen, but there are a few invertebrates that can survive under anaerobic conditions.

EXTERNAL AND INTERNAL RESPIRATION. Respiration involves a diffusion gradient of oxygen and carbon dioxide across a thin moist membrane, the oxygen being in greater concentration outside the cells, the carbon dioxide in greater concentration within the cells. Diffusion of these gases between the external environment and the animal constitutes *external respiration,* in some cases called *breathing.* It may occur across the walls of the alveoli or air sacs of lungs (air-breathing vertebrates), of lung books (spiders), of gills (fishes, mussels, larval amphibia), or of tracheal gills (in early stages of some insects, e.g., dragonflies). There may be no specialized structures, respiration occurring over the whole surface (*Hydra, Dugesia,* earthworm). The gas exchange with the individual cells is sometimes referred to as *internal respiration,* but this expression is somewhat ambiguous because cellular respiration is also internal respiration. In lower Metazoa, oxygen diffuses in from cell to cell; in insects, air is carried directly to the cells through tracheae; but in other higher animals (earthworm, mussel, all vertebrates), the blood-vascular system carries oxygen to the cells and carbon dioxide away. The gases may be in solution or may be carried in loose combination with a respiratory pigment, e.g., hemoglobin, a major oxygen carrier. Carbon dioxide is carried as carbonic acid and its ions, but the acid effect is typically buffered (see p. 16). In lower coelomate forms, having no circulatory system, the coelomic fluid may convey oxygen to the cells and carbon dioxide away.

Some animals consume oxygen at rates dependent upon the external oxygen concentration (*dependent respiration*), while others consume oxygen at a nearly constant rate even when the external oxygen concentration varies (*independent respiration*). The control of oxygen consumption is related to changes in oxidative metabolism, size of respiratory surfaces, rates of movement of air or water over respiratory surfaces, and the efficiency of transport by a circulatory system. Lack of oxygen may result in lactic acid accumulation, carbon dioxide increase, and internal increase in acidity. In mammals, cells sensitive to low oxygen, low pH, and high carbon dioxide activate the respiratory center, which stimulates stronger and more frequent breathing movements. Other reflex controls of respiratory movements involve blood pressure levels and the state of respiratory muscles.

MECHANICS OF BREATHING IN THE FROG. The mechanisms by which air is brought into the lungs vary considerably with different vertebrates. In the frog, breathing cannot occur with the mouth open, for it is by pressure within the mouth cavity, plus the opening of the glottis, that air is pushed into the lungs. The mechanism of breathing in man, which is quite different, is described in Chapter XV.

**Circulation.** The body fluids of some animals move (circulate) through the body in a more or less regular pattern. Such circulating fluids may have functions related to digestion, respiration, excretion, or endocrine regulation. If a circulatory system involves well-defined vessels and contractile components it is called a blood-vascular system, the contained fluid being called blood. In vertebrates, the tissue, coelomic, and cerebrospinal fluids are separated from the blood and have distinctive functional relationships.

BLOOD-VASCULAR SYSTEM. This involves a heart and a system of blood vessels. The system of vessels may be completely closed—arteries from the heart branching into capillaries, then reuniting in veins to return the blood (mussel, earthworm, vertebrates) ; or the heart may pump blood through arteries that do not merge into capillaries but empty their contents into large spaces through which the blood is drawn back to the heart (crayfish) ; or the blood may flow directly from the heart through irregular spaces in the coelom and back to the heart (insects). Differences between typical animals in the course of circulation are illustrated in Figures 11.7, 11.8, 11.12, 11.14, 11.20, 11.22, and 15.3. In vertebrates there is an evolutionary increase in the number of heart chambers: two in fishes, three in amphibians, three (and by some interpretations four) in reptiles, and four in birds and mammals. This increase in the number of heart chambers is accompanied by increasing separation of systemic and pulmonary circulations.

BLOOD. Blood consists of a fluid portion, the *plasma,* and, suspended in it, cells known as *corpuscles.* The dissolved constituents of the plasma and the corpuscles are all maintained in the blood at relatively constant levels. The inorganic constituents of the blood resemble in composition those of body fluids in general. Blood proteins may be distinctive, however, and of major importance; they serve in gas transport, act as buffers, bind and transport hormones, act as antibodies in immune reactions, are the initiators of and major constituents in blood coagulation, and influence the cell fluid osmotic relationships. Corpuscles of amoeboid form (*amoebocytes*) occur in a great variety of animals. In the vertebrates these are called "white corpuscles" or *leucocytes*. Some of them function, by phagocytosis, in the elimination of waste products and cell fragments; some are involved in immune reactions. Corpuscles of a second type, the "red corpuscles" or *erythrocytes,* also occur in vertebrates. These, of more stable form than leucocytes, contain the respiratory pigment, hemoglobin, and are involved in oxygen transport. There is also a third type of cellular component in vertebrates, the blood *platelets* or *thrombocytes,* which play a vital role in blood coagulation. These are usually cell fragments rather than complete cells.

**Excretion.** The waste products of metabolism are discharged from the body by excretion. (This is distinct from egestion, the discharge of materials that have not been absorbed in the digestive tract.) Small molecular end products of catabolism, such as ammonia and carbon dioxide, tend to diffuse from higher concentrations in cells into the extracellular body fluid. In many aquatic organisms (e.g., *Hydra*) the excretory products may simply diffuse from body fluids into the surrounding water. In flatworms, some annelids, rotifers, and *Amphioxus,* blind-ending excretory tubules (*protonephridia*) open directly to the outside. (These are also osmoregulatory.) At the inner end of the tubule are specialized cells from which extend one or many cilia (e.g., flame cells or bulbs of flatworms) into the lumen of the tubule. More advanced excretory tubules called *nephridia* drain the coelom directly (e.g., in earthworms). Secretory cells along these channels excrete some components of the coelomic fluid and return others, a functional pattern that is also present in higher animals. Pores opening from the coelom into the nephridia are still present in primitive vertebrates, but in higher vertebrates the excretory products are all derived from blood rather than from coelomic fluid. For details of the functioning of a nephridium in man, see Chapter XV. Higher vertebrates excrete nitrogen in the form of urea, which is less toxic in circulating body fluids than ammonia. The conversion of ammonia to urea takes place in the vertebrate liver in the *ornithine cycle.* The urea so formed contains nitrogen derived directly from amino groups as well as from ammonia. Reptiles, birds, and adult insects secrete uric acid, which can be concentrated and excreted in crystalline form—an adaptation that conserves water.

**Nerve Coordination.** The nervous system coordinates body functions in association with the generally slower, longer-term coordination by the endocrine system (see below). Through responses to environmental information and to internal changes, nerve impulses connecting various organ systems aid in body coordination. Furthermore, the nervous system develops action patterns independently of stimulation, these providing the basis for animal behavior (Chap. XXI).

STRUCTURAL BASIS OF THE NERVOUS SYSTEM. The *neuron* (nerve cell) is the fundamental unit of the nervous system. Neurons consist of a *cell body,* containing the nucleus, with branches or cell extensions that may be of considerable length. These branches have been distinguished on the basis of function: *dendrites,* branches that normally carry impulses toward the cell body, and *axons,* branches that normally carry impulses away from the cell body. However, this scheme has been replaced by some physiologists with a classification of more practical value, which is based on morphology, and which leads to somewhat different definitions of dendrites and axons. In this latter scheme dendrites are extensively branched areas, usually adjacent to the cell body, and axons are the generally single, slender, long processes that may possess specialized end structures. Side branches of axons that join other nerve cells are called *collaterals.* With axons, dendrites, and collaterals thus identified, physiologists recognize three functionally distinct regions in neurons: (1) the *generator region,* consisting of the dendrites, cell body, and generator collaterals; (2) the *conductile region,* the axons; and (3) the *transmission region,* nerve endings or terminals and transmission collaterals. All neurons have associations with other neurons. The *synapse* is the specialized region of contact between the transmission region of one neuron and the generator region of another. The transmission region may, alternatively, be associated with an effector (e.g., muscle) cell.

Nerve cells are surrounded by specialized cells called *glia cells.* Certain glia cells, the *Schwann cells* that surround some axons, give rise to the *myelin sheath.* Fluid exchange between the blood and cerebrospinal fluid in the vertebrate brain takes place through the physiologically active glia cells.

PROPERTIES OF IMPULSE CONDUCTION. The impulse, which passes

along a nerve as a result of a stimulus, moves at a measurable speed (seldom over 100 meters per second), which is variable from animal to animal and in different parts of the same animal. The impulse is accompanied by a change in electric potential that can be detected as a wave of negativity. This temporary change from the normal or *resting potential* (with the outer surface electrically positive) results in an *action potential* (with the outer surface negative). The change is caused by a shift in cell membrane permeability, which influences ion distributions ($Na^+$, $K^+$, and $Cl^-$ ions). Conduction is the propagation of this change along the axon. (It is therefore misleading to compare nerve conduction with conduction of an electric current along a wire.) The nerve has a *refractory period* or recovery stage following conduction, during which it is less sensitive and the resting potential is being restored through active transport. This phase is, of course, dependent upon energy derived from metabolism. The nerve impulse may pass in either direction along a neuron, but only from the transmission region of one neuron to the generator region of the next.

Transmission across the synapse (synaptic cleft) is probably mediated chiefly by secretions called *neurohumors,* though there are cases of electrical transmission. These neurohumors include *acetylcholine* and forms of *epinephrine.* They are released by the transmission region and influence the activity of the generator region, which initiates the action potential in the conductile region. A local enzyme immediately hydrolyzes the neurohumor, so it cannot accumulate. The relationships between neurons are often complex. Sometimes groups of impulses from one or more neurons are required to initiate an impulse. Furthermore, chemical production includes inhibitory as well as excitatory compounds. The delicate balance required between chemical production by transmission regions, sensitivity of generator regions, and the continual degradation of accumulating compounds is influenced by chemicals such as curare, strychnine, nerve gas, and some compounds used as pesticides.

THE MECHANISM OF A REFLEX ARC. A simple reflex arc involves at least: a *receptor* (sense organ), an *afferent neuron,* an *efferent neuron,* and an *effector* (motor organ). (See Fig. 15.5.) From a few to many *association* or *internuncial neurons* are nearly always present between afferent and efferent neurons. Most of the delay in a nervous response is due to the number of synapses rather than the total distance the impulse must travel.

**Sensation.** A nerve impulse is the result of a stimulus, often received by an external sense organ. As a result of specialized cell modifications and structural developments, the receptor neurons of a sense organ initiate impulses in response to specific environmental stimulations. The sensory system is the system that, more than any other, keeps the individual animal in contact with its environment.

SIGHT. In primitive organisms, sensitivity to light may occur in complete absence of specialized organs. In other animals, an organ that distinguishes between light and darkness may be present. In higher animals, an image from the outside world may be perceived by an eye. In all these cases the light-sensitive cells contain photosensitive pigments whose breakdown in light initiates nerve impulses.

The eye is best understood in analogy with a camera. In a camera, light is refracted by a *lens;* the eye also has a lens, but in addition two liquid refractive media, the *aqueous* and *vitreous humors.* One obtains sharp focus in a camera by moving the lens a greater or lesser distance from the sensitive plate; in primitive vertebrates, the lens may be moved a short distance within the eyeball, but in the human eye the shape itself of the lens is modified. An *iris* diaphragm is present in both camera and eye, concentrating the rays of light on the center of the lens. The *retina* of the vertebrate eye, composed of cells that contain light-sensitive chemicals, is equivalent to the sensitive plate of the camera. The analogy does not explain sight itself, however, for a camera cannot see. The animal sees with its brain centers, not its eyes; the eyes merely contain the centers for stimulation of the optic nerve tracts. (Fig. 15.6.)

EQUILIBRIUM AND HEARING. Organs of equilibrium occur in vertebrates, arthropods, and mollusks, at least. They all depend in function on movements of granules or liquids or both across sensitive nerve endings. In higher vertebrates, this function is associated with hearing, the ear housing both sense organs. (Fig. 15.7.) Organs of hearing present in arthropods as well as vertebrates include membranes that can be set in motion by alternate condensations and rarefactions of the air. The impulse thus set up is conducted by an intermediary, e.g., the chain of ossicles in the human ear, to the real end organ of hearing, which contains the nerve endings. Ability to distinguish between sounds of different pitch involves a very complicated end organ, which, according to some theories, responds by sympathetic vibration to different frequencies at different points along its surface. One interesting adaptation of a sense of hearing is the "sonar" of bats. These animals emit ultrasonic sounds, whose echos they detect in flight and use in avoiding obstructions or in catching insects. Some insects (moths) in turn have developed the ability to detect ultrasonic sounds, which enables them to avoid their predators.

CHEMICAL SENSES. The senses of smell and taste involve the exposure of nerve endings to volatile and nonvolatile materials respectively, in solution. Such sense organs occur only in moist membranes—e.g., frog skin.

TACTILE SENSES. End organs of a corpuscular nature, detecting sensations of pressure, pain, heat, and cold, occur in the skin and various body membranes.

PROPRIOCEPTORS. Sense organs similar to the last type in structure occur in muscles and tendons. Impulses set up in them during the contraction of muscles help maintain reciprocal action of antagonistic muscles—preventing muscles of opposite effects from working against each other.

**Chemical Coordination—Endocrine Secretion.** In the higher vertebrates, and less strikingly in other forms (arthropods, in particular), there occur glands that empty their secretions directly into the bloodstream. These secretions are carried in the blood to all parts of the body, and they influence activities at a distance from their place of origin. The glands are called *endocrine glands,* or glands of internal secretion. Their products are called *hormones* ("chemical messengers"). In the arthropods, they induce color changes and are influential in regulating development. Insect metamorphosis is under hormonal control. In vertebrates, hormones control a variety of metabolic, developmental, and reproductive functions. In frogs, for example, thyroid secretion stimulates the metamorphosis from tadpole to adult frog. The endocrine glands of man, and the functions of the hormones secreted by them, both better known than those of other animals, are summarized in Chapter XV.

# Chapter XIII

# REPRODUCTION, EMBRYOLOGY, AND DEVELOPMENT OF MULTICELLULAR ANIMALS

Reproduction in multicellular animals may occur by either sexual or asexual means; in the higher animals it is exclusively sexual. The two methods differ in that asexual reproduction is uniparental and only mitotic divisions are involved between parent and progeny whereas sexual reproduction is biparental and is accompanied by meiosis and fertilization. (See Chap. X.)

## ASEXUAL REPRODUCTION

In Metazoa, asexual reproduction involves development of offspring from a relatively small part of the parent—not equal division of parent into two offspring, as in Protozoa.

**Budding.** In sponges, coelenterates, and certain other invertebrates, new individuals may form as outgrowths from the parent. They may later become separate individuals (*Hydra*) or remain attached as parts of a colony (*Obelia*). In cestodes (tapeworms) new individuals (proglottids) bud from the neck, but remain attached together, distal proglottids dropping off when mature. Certain annelids divide transversely into individuals which separate, becoming independent—a form of reproduction intermediate between budding and fragmentation; a similar method occurs among planarians.

**Fragmentation.** Some planarians break into fragments, each of which reorganizes into a new individual.

**Gemmule Formation.** In certain sponges, particularly freshwater forms, at the onset of unfavorable weather, germinal cells become grouped into balls surrounded by spicules. The parent sponge dies, but the balls or *gemmules* survive the unfavorable period, developing into sponges during the next season.

**Statoblast Formation.** Freshwater ectoprocts form disc-shaped resistant structures, *statoblasts,* which after an unfavorable period (which may have killed the parent colony) germinate into new colonies.

# SEXUAL REPRODUCTION

In normal sexual reproduction two mature germ cells (gametes) fuse to form a *zygote.* The gametes have developed in parents of different *sex.* The male gamete is a *sperm cell* (*spermatozoon*); the female, an *egg cell* (*ovum*). By a process called *maturation* or *gametogenesis,* each gamete develops within an organ called a *gonad.* The gonad of the male is the *testis;* of the female, the *ovary.* The fusion of the gametes, particularly of their nuclei, constitutes *fertilization.* The zygote becomes the mature animal through the process of *embryogeny.*

**Variations from Normal Sexual Reproduction.**

HERMAPHRODITISM. One individual, a *hermaphrodite,* may produce both eggs and sperm cells. The eggs may be fertilized by sperm cells of the same animal (*self-fertilization*) or by those of another (*cross-fertilization*). (Examples: *Hydra,* flatworm, earthworm, snail.)

PARTHENOGENESIS. This is reproduction through the development of an unfertilized egg—so-called virgin reproduction. (Examples: bees, plant lice. *Artificial parthenogenesis* has been induced by chemical or physical methods in many animals.)

PAEDOGENESIS. Reproduction by an animal in a young or larval condition is called *paedogenesis.* (Examples: liver fluke, tiger salamander.)

**Gametogenesis.** This is the process by which gametes are formed. Structurally, the male gamete is modified for active movement and for penetration of the egg. The female gamete contains a large amount of food material (yolk) for the developing embryo and is passive in movement. In both, the chromosome number has been reduced from diploid to haploid. (Fig. 13.1.)

SPERMATOGENESIS. Following the period of cellular multiplication which increases the number of germ cells (*spermatogonia*) in the testis, there occurs the growth period. The growing cell (destined to form spermatozoa) is the *primary spermatocyte.* It divides, forming two *secondary spermatocytes.* Each of these divides, forming two *spermatids.* Each spermatid, by a process of differentiation without any division,

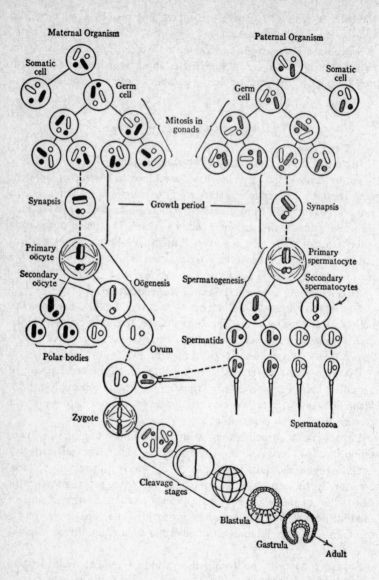

Fig. 13.1. Diagram illustrating gametogenesis (oögenesis in the female, spermatogenesis in the male) and the early stages of embryonic development in animals. Also illustrated is the continuity of chromosomes from the zygote that constitutes each parent, through the germ cells in the gonads of these individuals, to the embryo that is their progeny.

becomes a sperm cell or spermatozoon. Two cell divisions have occurred, together constituting *meiosis*. End result: four functional sperm cells, each with the haploid (*n*) chromosome number.

OÖGENESIS. Following the period of multiplication of germ cells (*oögonia*) in the ovary, there occurs the growth period, the enlarged cell being the *primary oöcyte*. It divides unequally, forming one (large) *secondary oöcyte* and a (small) *polar body*. The size difference is not due to nuclear size difference, but to cytoplasmic. The polar body may or may not divide again; its fate is unimportant. The secondary oöcyte divides to form a second small polar body and the mature egg cell or ovum. The divisions result in one ovum with the haploid (*n*) chromosome number, and two or three nonfunctional polar bodies.

**Fertilization.** Fertilization of the egg is completed when the two *pronuclei,* one from the sperm cell, one already in the ovum, fuse. The sperm may penetrate the egg cell before the first polar body is formed, between the formation of first and second polar bodies, or after the egg is mature; in the first two cases it does not fuse with the egg pronucleus until maturation is complete. The time of *penetration* is characteristic for different species. Fertilization merges immediately into the first division of the zygote, the chromosomes from the two pronuclei being combined on the mitotic spindle, restoring the diploid (2*n*) chromosome number and giving to the new individual equal numbers of chromosomes from paternal and maternal parents. (Fig. 13.1.)

In this process the nuclear contributions of the two gametes are equal, but the cytoplasmic contributions are not. The cytoplasm of the egg typically contains both organelles and stored food that are not contributed by the male gamete. In parthenogenesis, on the other hand, both nucleus and cytoplasm are solely from the female gamete. The developing cell sometimes gives rise to a diploid chromosome complement by doubling of its own haploid set; in other cases the individual produced by parthenogenesis is haploid.

**Accessory Processes.** Various processes adapted to insure fertilization occur in different species. If fertilization occurs outside the female's body, both egg and sperm cells may be shed into the surrounding medium (in every case, water). In some cases where fertilization is external, behavior patterns exist which insure a greater percentage of fertile eggs: in the earthworm, eggs are discharged into a cocoon, into which are then ejected sperm cells received previously during copulation with another worm; in frogs, a male clasps a female during spawning, so that the sperm cells as discharged come immediately into contact with the egg mass. In animals in which fertilization is internal, the male discharges the sperm cells directly into the genital tract of the female during *copulation.* In some forms, this involves merely the apposition of the genital openings, neither being specialized. In salamanders, the male discharges packets of sperm cells which are later picked up by the cloaca of the female. An *intromittent organ* or *penis,* introduced into the female genital tract during copulation, may convey packets of sperm cells (grasshopper) or sperm cells in a liquid medium produced by various accessory glands (higher vertebrates).

## EMBRYOLOGY OF METAZOA

The whole course of development from the fertilized egg is sometimes called embryogeny. Ordinarily, however, embryogeny is considered the process of development from the fertilized egg to a stage approximating the adult condition, e.g., to hatching in birds or birth in mammals.

Embryology is the study of the process of embryo development.

**General Principles.** The zygote divides by a sequence of mitoses characteristic for each animal, this constituting a differentiation pattern that produces the adult form. The cells first divide synchronously, but soon those in one part of the embryo are dividing independently of others, resulting in variations in shape and size of different parts. Such differences in patterns of cell division pave the way for differentiation into organs and for development of the differences among organisms.

**Kinds of Egg Cells and Types of Cleavage.** The nucleus of egg cells is usually eccentric, its position being near the side on which the polar bodies are formed—the *animal pole.* The cytoplasm is concentrated toward the same side, and the yolk toward the opposite side—the *vegetal pole.* If the yolk is small in amount and rather evenly distributed through the egg (as in the eggs of many invertebrates, *Amphioxus,* and placental mammals), the egg is *isolecithal.* When a fertilized egg of this kind divides, its cleavage is complete, and the daughter cells are usually of about the same size. Complete division is called *holoblastic cleavage.* If the yolk is large in amount and concentrated toward the vegetal pole (as in fish, amphibia, reptiles, and birds), the egg is *telolecithal.* Some telolecithal eggs (e.g., those of the frog) undergo holoblastic cleavage, with the daughter cells quite unequal in size, those at the vegetal pole being much larger than those at the animal pole. Telolecithal eggs with the largest amounts of yolk (in reptiles and birds) divide only in a disclike region at the animal pole, this process being called *meroblastic cleavage*. In insect eggs, the yolk, which is central, is surrounded by a layer of cytoplasm. Such an egg is *centrolecithal,* and its cleavage occurs on the surface, all over the egg.

As the fertilized egg divides the various regions of the cytoplasm become segregated, and these eventually give rise to different parts of the embryo. If the destiny of these different regions is established early, so that separation of the first cells results in development of only partial embryos, the egg is said to be *mosaic* and to have *determined development*. If the cells of the early embryo can be separated with each developing into a whole embryo, the egg is *regulative* and it has *undetermined development*. Many invertebrates have mosaic eggs, while vertebrates have regulative eggs. As embryonic development proceeds the cells of regulative types become essentially mosaic at later stages.

**Stages in Embryogeny.** The following stages merge one into another; they are given different names for convenience only. The basic pattern summarized here is that associated with holoblastic cleavage as it occurs in the sea urchin or *Amphioxus*. Modifications are described as variations from this pattern. (See Figs. 13.1 and 13.2.)

EARLY CLEAVAGE. The cells as they divide remain attached; the original mass does not increase in size, it simply subdivides. The zygote forms two cells, each of these two making four; in some cases this geometric progression continues for some time: 2—4—8—16—32; but in most cases, greater division rates at the animal pole disturb the regularity. Embryos of approximately sixteen cells look like mulberries; hence the name *morula* for that stage.

BLASTULA. Continued cleavage results in the formation of a hollow ball of cells, the *blastula,* its cavity being the *blastocoel*.

GASTRULA. More rapid division of cells at the animal than at the vegetal pole results in a pushing in or *invagination* of the cells near the vegetal pole. This results in a reduction of the blastocoel as the outer cells push inward, the embryo becoming a ball of cells whose cavity (*archenteron*) opens to the outside but is surrounded elsewhere than at the opening (the *blastopore*) by two layers of cells, the outer *ectoderm* and the inner *endoderm* (Fig. 13.1). During gastrulation in the frog, invagination is accompanied by *involution,* a process in which outer cells roll under the edge of the blastopore (Fig. 13.2*I, J, K*). In animals with eggs having a large amount of yolk (e.g., birds) invagination does not occur, but the endoderm is formed by *delamination*—an internal

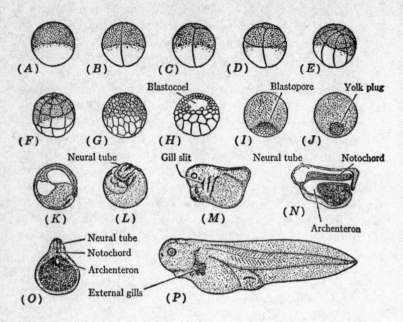

Fig. 13.2. Stages in the embryology of the frog. (*A*) The uncleaved egg. (*B-F*) Early cleavage stages. (*G-H*) Blastula. (*I-K*) Gastrula. (*L-P*) Organ formation. (*N*) Longitudinal section and (*O*) cross section of embryo shown at (*M*). (Adapted from figure copyrighted by General Biological Supply House, Chicago, and used by permission.)

layering-off from the cells of the blastula. (The blastula, in this case, is constituted by a superficial layer of cells.) A thickened region of contact between ectoderm and endoderm, the *primitive streak,* is homologous with the blastopore. The internal cavity of such embryos becomes secondarily open to the outside. From an evolutionary standpoint, the gastrula represents the stage to which an adult coelenterate attains.

MESODERM FORMATION. Between ectoderm and endoderm, cells proliferate to occupy the segmentation cavity. These form the third germ layer, the *mesoderm.* The platyhelminthes attain this stage in evolution but go no further.

COELOM FORMATION. Mesoderm cells may separate into two layers, an outer (*somatic*) and an inner (*splanchnic*), leaving a space between which becomes the true *body cavity* or *coelom.* The coelom may be separated by transverse partitions into segmentally arranged cavities (e.g., earthworm).

ORGANOGENY. With the beginning of coelom formation, *organogeny* or differentiation of organs commences. (Example: formation of nervous system from ridge of ectoderm.) Derivatives of the embryonic germ layers are, in general, as follows:

From *ectoderm:* Integument, nervous system, lining of mouth and anus.

From *endoderm:* Lining of alimentary canal (except mouth and anus) and its extensions (e.g., lungs, bile ducts).

From *mesoderm:* Internal skeleton, muscles, heart and blood vessels, kidneys, gonads, muscular and connective tissue layers of alimentary canal and its extensions.

**Embryonic Induction.** The axis of the embryo is determined by cells at the dorsal lip of the blastopore, the *organizer.* This group of cells illustrates the general principle of *embryonic induction.* Throughout the period of organ development certain cells play a role in determining the fate of adjacent cells. Cell differentiation is not, however, the only process involved. The death and absorption of certain cells may be involved if structures are destined to disappear (e.g., the tadpole tail) or be replaced (e.g., membrane by bone).

**Accessory Structures.** The embryos of many animals (e.g., many invertebrates, fish, amphibia) develop, without special protection, in natural freshwater or marine habitats. In some terrestrial animals (e.g., insects, reptiles, birds) the embryo develops outside the body but within an egg membrane or shell that protects it from desiccation. We call such types of development *oviparous.* In other, mainly terrestrial animals the embryo is retained during a period of development in an expanded portion of the oviduct, the *uterus.* In some such cases (e.g., live-bearing snakes), the egg is merely protected, being retained in the uterus during incubation (*ovoviviparous* development). In others, however (e.g., placental mammals), the embryo receives nourishment from the mother. This is *viviparous* development.

Whether within a uterus or not, the embryos of vertebrates develop special *embryonic membranes* that aid in nutrition, respiration, excretion, or protection (Fig. 13.3). In lower vertebrates, a *yolk sac* develops as a diverticulum from the embryonic gut, enclosing the food supply of the embryo. In reptiles, birds, and mammals, there are three embryonic membranes in addition to the yolk sac: (1) the *amnion,* nearest the body wall of the embryo; (2) the *chorion,* outside the amnion; and (3) the *allantois,* which develops as a diverticulum of the hindgut that grows out between amnion and chorion. The amnion and chorion are

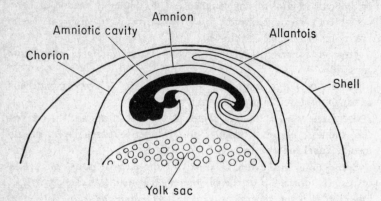

Fig. 13.3. Diagram illustrating the relations of the embryonic membranes in a developing bird embryo. The chorion, amnion, allantois, and walls of the yolk sac are all double layers, but they are here shown, for simplicity, with single lines. The body of the embryo, in longitudinal section, is indicated by the solid portion. (Reprinted by permission from *General Biology* by Gordon Alexander, published by Thomas Y. Crowell Company.)

of ectoderm and mesoderm, the yolk sac and allantois of endoderm and mesoderm. The amnion encloses a fluid-filled cavity, the *amniotic cavity,* which surrounds the embryo. In mammals, the allantois and chorion, in association with the lining of the uterus, typically form the *placenta,* an organ through which exchanges between fetal (embryonic) and maternal circulatory systems take place. The mesoderm of the allantois and yolk sac contributes blood vessels, which transport nutrients from the yolk and placenta and respiratory gases to and from the lunglike tissues within the egg or the placenta. Food and oxygen may diffuse through the placental membranes from the maternal blood into the fetal circulation and waste materials may diffuse in the opposite direction, but there is no continuity between the two circulatory systems. The *umbilical cord* carries the blood vessels that connect embryo and placenta.

## COMPLEXITIES IN ANIMAL DEVELOPMENT

**Metagenesis.** Alternation of a sexually reproducing with an asexually reproducing generation occurs in some animals. It is best illustrated in colonial coelenterates like *Obelia* where it is called *metagenesis.* The animals of the colony are formed asexually, by budding; certain of these buds produce, also by budding, the medusae; the medusae reproduce sexually, the fertilized eggs developing into polyps which, by budding, produce new colonies. (Fig. 11.4.) Metagenesis also occurs among the parasitic flatworms: (1) the adult sheep liver fluke reproduces sexually; from the egg develops a ciliated larva which bores into a snail; inside, it becomes a *sporocyst* which produces *rediae* asexually; these produce more rediae, and then *cercariae,* asexually; from the cercariae adults develop. (2) In the tapeworm life history, the proglottids are sexually reproducing individuals, but these have been derived from the scolex by an asexual method, viz., budding. In all these cases, whether sexual or asexual reproduction is involved, the organisms are diploid. This process is not homologous with alternation of generations in plants (Chap. X).

**Indirect Development.** Many animals attain adult form only after passing through a series of stages more or less unlike the adult. If these stages are very unlike the adult, development is said to be indirect (*complete metamorphosis*); if these stages are present but quite like the adult, in general, development is said to be direct (*incomplete metamorphosis*). Common examples of the former are (1) frogs and toads —which pass through a tadpole stage and (2) bees and butterflies— which have larval and pupal stages. (Among insects having complete metamorphosis, four stages may be recognized: egg, larva—during which stage there may be several molts, pupa, adult.) A common example of incomplete metamorphosis is furnished by the grasshopper, which passes through a series of nymphal stages all of which look somewhat like the adult. The differences are in body proportions and presence and nature of wings or wingpads.

*Part Three*

# MAN: MORPHOLOGY, PHYSIOLOGY, AND REPRODUCTION

# Chapter XIV

# MAN: PROTECTION, SUPPORT, MOVEMENT

Man is an animal. As an animal, he belongs to the species *Homo sapiens* of the order Primates, class Mammalia, phylum Chordata. His anatomy and physiology lend themselves to the same kind of study carried out on other animals. Man is distinct from other animals in the possession of conceptual thought and speech; and, in connection with these, he has a highly organized social system and such distinctly human developments as government and law, education, and religion. Man is, therefore, a unique creature, but the biologists' concern with him is not in his uniqueness but in the ways in which he is a typical animal. In this and the next two chapters each organ system will be considered from the twin aspects of morphology and physiology.

## GENERAL CHARACTERISTICS

**Body Divisions.** The major divisions of the human body are the *head, neck, trunk,* and two pairs of appendages, the *arms* and *legs*. Five *digits* are present on each appendage. The trunk is commonly divided into two regions, the *thorax,* containing the heart and lungs, and the *abdomen,* containing the liver, stomach, intestine, and other viscera. The axis of the head is horizontal—at right angles to the perpendicular axis of the body. The ventral surface of the trunk is in front, the dorsal surface, in back. The head is anterior to the trunk, not dorsal to it.

**Coelom.** The coelom, in man and other mammals, has four compartments. These are the *pericardial cavity* (containing the heart), two *pleural cavities* (each containing a lung), and the *peritoneal cavity* (containing the abdominal viscera). The peritoneal cavity is bounded anteriorly by the *diaphragm,* a transverse muscle.

# INTEGUMENTARY SYSTEM

The *skin* consists of two layers, an outer thin *epidermis* and an inner thick *corium* (or *dermis*). The inner layer, but not the epidermis, contains a variety of tissue types, including blood vessels and nerves. The epidermis, of epithelial tissue, protects underlying structures from loss of moisture and from injury. Specialized structures of the integument include *hair, nails,* and *glands.* The hair of man is distributed on the head, in the axillae (armpits), and in the pubic region in both sexes, and has a somewhat wider distribution in the male, including the face. Hair and finger and toe nails are derived from the epidermis. The principal glands in the skin are *sweat glands,* which function in temperature regulation (evaporation being a cooling process) and in excretion. Other integumentary glands include *oil glands* and, in the female, *mammary* or *milk glands.* Hair and mammary glands are characteristics man shares with all other mammals.

# SKELETAL SYSTEM

**Morphology.** The bones of the human skeleton comprise an *axial* and an *appendicular skeleton.* The former includes the *skull, vertebral column, ribs,* and *sternum;* and the appendicular skeleton comprises the bones of the appendages and *girdles.* All bones of the human skeleton are summarized in Table 14.1, and their general relations are indicated in Figure 14.1.

**Physiology.** The major functions of the skeleton are support, protection, and movement (supplying points of attachment for the skeletal muscles). The normal formation of bone in a developing child depends upon the presence of a dietary element, vitamin D, contained plentifully in cod-liver oil and capable of being formed in one's body on exposure to sunlight. The ratio of calcium in blood and bone is regulated by the hormone secreted by the parathyroid glands. A growth-regulating hormone from the anterior lobe of the pituitary gland is involved in bone growth; its excessive secretion may cause gigantism; inadequate secretion may be associated with dwarfism.

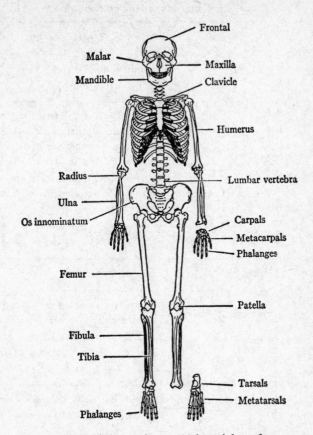

Fig. 14.1. The skeleton of man. (Adapted from figure copy-righted by General Biological Supply House, Chicago, and used by permission.)

## MUSCULAR SYSTEM

**Morphology.** A skeletal muscle consists typically of a mass of striated muscle fibers extending across a joint of the skeleton. The end that moves least during contraction is the *origin;* the end that moves most, the *insertion.* The movement accomplished constitutes the *action.* Table 14.2 gives origins, insertions, and actions of a few typical skeletal muscles. Some striated muscles are not associated with the skeleton. The diaphragm has a circular origin; its insertion is a central tendon. The tongue is a mass of striated muscle fibers extending in three planes.

**Table 14.1. Bones of the Human Skeleton**

| | | | |
|---|---|---|---|
| Axial Skeleton | Skull | Cranium | 1 occipital, 2 parietals, 1 frontal, 2 temporals (each containing 3 ossicles of middle ear), 1 sphenoid, 1 ethmoid. |
| | | Face | 2 nasals, 1 vomer, 2 inferior turbinals, 2 lacrimals, 2 malars, 2 palatines, 2 maxillae, 1 mandible, 1 hyoid. |
| | Vertebral Column | | Vertebrae: 7 cervical, 12 thoracic, 5 lumbar, 5 sacral (fused into 1), 1+ coccygeal. |
| | Thoracic Basket | | 24 ribs, 1 sternum (of 3 elements). |
| Appendicular Skeleton * | Pectoral Girdle | | 1 clavicle, 1 scapula. |
| | Pectoral Appendage | | 1 humerus, 1 radius, 1 ulna, 8 carpals, 5 metacarpals, 14 phalanges. |
| | Pelvic Girdle | | 1 os innominatum, consisting of ilium, ischium, and pubis. |
| | Pelvic Appendage | | 1 femur, 1 patella, 1 tibia, 1 fibula, 7 tarsals, 5 metatarsals, 14 phalanges. |

* Bones of the appendicular skeleton are given from only one side; to arrive at the total number of bones each of these figures should be multiplied by two.

Nonstriated, *smooth muscle,* and *cardiac muscle,* which is striated, are not ordinarily considered parts of the muscular system because their major activities are related to functions other than locomotion.

**Physiology.** Muscular contraction involves the shortening of muscle fibers without any change in volume. Hence they become thicker as they become shorter. The shortening is due to microscopic changes in the molecules present in the myofibrils.

MECHANICAL RELATIONS. The shortening of skeletal muscles usually takes place across a joint, which, if bent, is said to be flexed, if straightened is said to be extended. The muscles involved are called *flexors* and *extensors*, respectively, and the flexors and extensors at a particular joint are *antagonistic muscles* (compare the actions of biceps and triceps in Table 14.2). The action of a muscle across a joint may be analyzed as one of the three classes of levers, the fulcrum being the joint, the power being applied at the insertion of the muscle, and the weight being the center of gravity of the movable part. All three classes occur in the human body (Fig. 14.2), third-class levers being the most frequent. The actions of the diaphragm and tongue, though these organs are made up of skeletal muscle, do not depend upon skeletal relations. The former is conical when relaxed; contraction transforms the cone into a flattened disc, increasing the vertical diameter of the thoracic cavity.

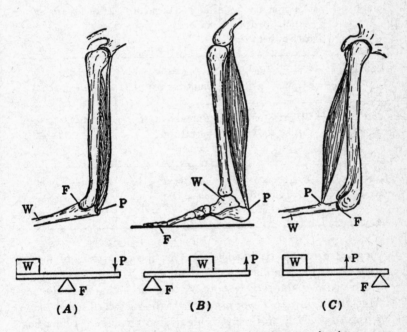

Fig. 14.2. Diagrams to suggest the relations between the three classes of levers and the actions of skeletal muscles. (*A*) First-class lever, illustrated by the triceps muscle extending the arm at the elbow. (*B*) Second-class lever, illustrated by the gastrocnemius muscle raising the weight of the body on the toes. (*C*) Third-class lever, illustrated by the biceps muscle flexing the arm at the elbow. Abbreviations: *F*, fulcrum, *P*, power, *W*, weight.

**Table 14.2. Origins, Insertions, and Actions of Representative
Muscles of the Human Body**

| NAME OF MUSCLE | ORIGIN | INSERTION | ACTION |
|---|---|---|---|
| Masseter | Zygomatic arch; temporal and malar bones | Mandible, at angle of jaw | Closes jaws |
| Biceps | Two points on scapula | Tubercle on radius | Flexes arm at elbow |
| Triceps | One point on scapula; two on back of humerus | Proximal end of ulna | Extends arm at elbow |
| Deltoid | Clavicle and scapula | Outer surface of humerus | Raises arm outward from body |
| Pectoralis major | Clavicle, sternum, cartilages of ribs | Ridge on outer surface of humerus | Pulls arm forward |
| Gastrocnemius | Two spots on distal end of femur at back | Calcaneum—the tarsal bone of the heel | Extends foot at ankle |

Tongue movements are brought about by the contraction of muscle
groups that extend in different directions.

CHEMISTRY OF MUSCULAR ACTIVITY. The chemical processes in muscle contraction involve a type of trigger reaction in which stored energy
is released suddenly, this energy being used to produce the contraction.
The contraction involves the muscle proteins, *actin* and *myosin,* present
in the *myofibrils* of the muscle cells. When a source of energy is released,
these proteins combine as *actomyosin,* resulting in a shortening of the
myofibrils. The energy is released by the breakdown of ATP and phosphocreatine. Both are common constituents of muscle. The ATP and
phosphocreatine are then resynthesized, the energy for this coming from
the oxidation of sugar or the formation of lactic acid from glycogen

(the latter process being anaerobic). Glycogen is the form in which carbohydrate food is stored in muscle. Most of the lactic acid may later be reconverted into glycogen, the energy for that process coming from the aerobic oxidation of a part of the lactic acid. Thus only the recovery phase involves oxygen, and a muscle can contract vigorously for some time in the absence of oxygen. A muscle becomes fatigued, however, and ceases to contract, if contractions occur so frequently that oxidative recovery cannot take place. (For more details see Chapter XII.)

# Chapter XV

# MAN: METABOLISM AND IRRITABILITY

## DIGESTIVE SYSTEM

Metabolic activity depends upon food as a source of energy. Man's specific needs in this respect constitute proper *nutrition*. The digestive process is basic to all other metabolic activities because man, like other animals, cannot manufacture his own food. He acquires it in complex form from plants (either directly or indirectly—from other animals) and, in order to use it, must reduce it to chemically simpler form. The process of preparing food for absorption and use is the function of the digestive system. The relations between its functions and those of other metabolic systems are illustrated in Figure 15.1.

**Morphology.**

ALIMENTARY CANAL. The digestive tract or alimentary canal consists, in order, of the *mouth, pharynx, esophagus, stomach, small intestine, large intestine* or *colon, rectum,* and *anus*. The mouth is closed by fleshy lips, and these and the tongue function in moving food so that the masticating organs, the *teeth,* can break it up. There are thirty-two teeth, of four types: eight *incisors,* four *canines,* eight *premolars,* and twelve *molars,* equally distributed in upper and lower jaws. These constitute the *permanent dentition*. The first or *milk dentition* lacks the twelve molars. The tongue bears *taste buds*. The mouth merges into the pharynx, the region in which air passing to and from the lungs crosses the food passage. The esophagus is a straight, muscular tube opening into the stomach. The latter is curved, its *greater curvature* to the left. The upper end of the stomach is the *cardiac end;* the lower, the *pyloric end,* terminates in a circular muscle, the *pyloric valve*. The small intestine consists of three parts, the *duodenum, jejunum,* and *ileum*. The large intestine or colon consists of ascending, transverse, and descending portions. A *caecum* (blind pouch) is present at the junction of colon

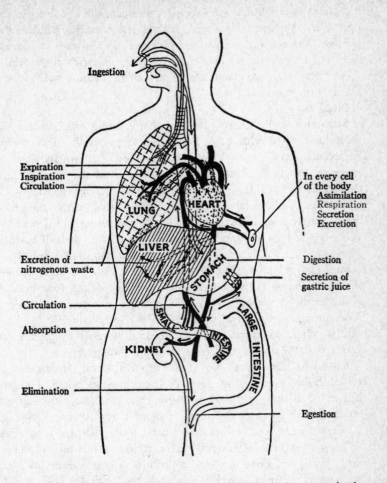

Fig. 15.1. Diagram to suggest the steps in metabolism as they occur in the human body. (Reprinted by permission from *Animal Biology* by R. H. Wolcott, published by the McGraw-Hill Book Company.)

and ileum, and the *vermiform appendix* is a projection from the caecum. The descending portion of the colon merges into the *sigmoid flexure* and rectum, the latter opening to the outside through the anus. There is no cloaca.

DIGESTIVE GLANDS. Three pairs of *salivary glands* empty into the mouth. The specialized digestive glands of the peritoneal cavity are the *liver* and *pancreas*. The liver, which is very large, consists of four lobes;

it contains a *gall bladder* (in which the secretion *bile* is stored) in the right lobe. The pancreas is a smaller, transverse organ lying behind the stomach. The *bile duct* (from the liver) and the *pancreatic duct* meet, their contents entering the duodenum together, not far from the pyloric valve. Other digestive glands are present in the stomach wall and the lining of the small intestine.

## Physiology.

MECHANICAL PROCESSES. The food, after being masticated in the mouth and mixed with *saliva* from the salivary glands, is swallowed by voluntary movements. Its passage through the alimentary canal is, however, largely involuntary. Food is retained in the stomach until thoroughly acidified. It is passed into the duodenum through the pyloric valve a small amount at a time. The acidified food in the duodenum causes the secretion of two hormones, one stimulating secretion of the pancreas and the other causing the flow of bile from the gall bladder. Mechanical actions of the intestine involve forward movement of the contents by *peristalsis* (muscle waves) and a rhythmic segmentation, the latter process allowing for complete mixing of the food mass or *chyme* with intestinal enzymes and complete absorption of digested food through the intestinal walls. Water is resorbed through the walls of the colon, concentrating the undigested material to form *feces*. *Defecation,* the discharge of fecal material, is voluntary.

CHEMICAL PROCESSES—THE DIGESTIVE ENZYMES. Digestion is accomplished by a series of specific hydrolyzing enzymes. The first enzyme activity is in the mouth, where *salivary amylase* (*diastase*) converts starch to maltose (a double sugar), but its action is stopped by high acidity when the food reaches the stomach. In the stomach, *pepsinogen,* an enzyme precursor, is converted to the protease *pepsin* by hydrochloric acid; pepsin hydrolyzes proteins to peptides. The pancreatic juice contains several enzymes, as well as substances that neutralize acid from the stomach. One of the enzymes is *pancreatic amylase,* with the same function as salivary amylase. Another is *lipase,* which hydrolyzes fats to glycerol and fatty acids. In addition to these, the pancreas secretes the precursor *trypsinogen,* which is converted to *trypsin,* a protease, by *enterokinase,* a secretion of the intestine. Trypsin digests proteins to peptides. Nucleic acids are also digested (to nucleotides) in the small intestine by enzymes secreted by the pancreas. Peptides are digested to amino acids by enzymes secreted by gland cells in the wall of the small intestine. Other enzymes secreted by the small intestine include those that split the common disaccharides (including maltose) into simple sugars. The bile does not contain a digestive enzyme but does aid in fat digestion, emulsifying fat particles.

FOOD ABSORPTION. The small intestine contains in its lining minute fingerlike projections called *villi* (sing., *villus*). These contain microscopic branches of the circulatory system—both blood capillaries and *lacteals* (lymphatics). In general, the end products of carbohydrate and protein digestion are absorbed in the blood capillaries and fat in the lacteals. The blood capillaries carry food to the liver in the portal system; the lymphatic system empties directly into the veins near the heart.

**Nutrition.** Nutritional requirements include food for energy, specific foods, water and mineral needs, and vitamins.

ENERGY REQUIREMENTS. Man's food requirements for energy are measured in *Calories* per day. (A Calorie, the unit of energy used in the description of human nutritional requirements, is 1,000 calories—a kilocalorie. Spelled with a capital "C" it signifies an amount of energy equivalent to 1,000 times that of the calorie ordinarily used in physics, which is the amount of heat energy necessary to raise the temperature of a gram of water from 15° to 16°C.) *Total metabolism* requires 2,500 to 5,000 Calories per day, depending upon the kind of physical exertion involved. *Basal metabolism,* energy use during complete rest, involves 1,600 to 1,800 Calories per day. Carbohydrates and proteins yield in human metabolism four Calories per gram, fats nine Calories per gram.

SPECIFIC FOOD REQUIREMENTS. A minimum protein intake is necessary to offset continuous nitrogen excretion—to maintain *nitrogen equilibrium*. Specific amino acids and fatty acids are required, too, particularly in growing individuals.

WATER AND MINERAL REQUIREMENTS. A human being requires about one quart of water per day, in addition to water contained in food. Most mineral requirements are present in an average diet in sufficient quantity, but phosphorus, calcium, and iron are three important elements in which a diet may be deficient.

VITAMINS. These are complex substances of plant origin required in man's diet but not needed in large quantity. Water-soluble vitamins include those of the vitamin B complex (thiamine, riboflavin, nicotinic acid, pantothenic acid, and others). These are components of coenzymes involved in cellular metabolism. Vitamin C (ascorbic acid) also has a role in cellular oxidations. The fat-soluble vitamins include vitamin A, required for normal maintenance of epithelial and certain other types of tissue and involved in the chemistry of vision; vitamin D, which man can synthesize in sunlight, related to bone formation; vitamin E, which may be related to sterility; and vitamin K, involved in blood clotting. Certain diseases (called *deficiency diseases*) occur in the absence of an adequate quantity of particular vitamins in the diet. Thus, scurvy is a result of inadequate vitamin C, and rickets is due to a deficiency of vitamin D.

# CIRCULATORY SYSTEM

**Morphology.** The circulatory system consists of a circulating medium, the *blood,* the channels through which it flows, and a *lymphatic system.* (Fig. 15.2.)

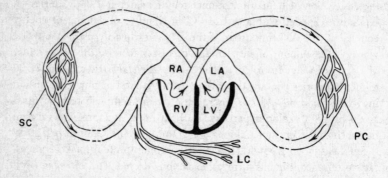

Fig. 15.2. Diagram illustrating the circulatory relations of capillary channels, major vessels, and heart chambers in man. Capillary systems: *LC,* lymphatic capillaries, which are distributed throughout the body and which are one-way channels emptying into veins near the heart; *PC,* pulmonary blood capillaries, surrounding the air sacs of the lungs; *SC,* systemic blood capillaries—i.e., all blood capillaries not in the pulmonary circulation. Heart chambers: *LA,* left atrium; *RA,* right atrium; *LV,* left ventricle; *RV,* right ventricle.

BLOOD. The blood consists of a liquid substrate, the *plasma,* in which are contained *erythrocytes (red corpuscles), leucocytes (white corpuscles),* and *thrombocytes (blood platelets).*

HEART AND BLOOD VESSELS. Circulation of the blood is maintained by the rhythmic contraction of the *heart,* the rate of the heart beat being intrinsic but subject to modification by autonomic nervous control. There are two *auricles* or *atria* in the heart, both thin-walled, and two *ventricles,* the latter with thick, muscular walls. Valves maintain movement of blood through the heart in only one direction. Blood to the lungs leaves the heart from the *right ventricle* by way of the *pulmonary artery,* whose opening from the ventricle is guarded by three *semilunar valves.* This artery divides into two pulmonary arteries, one to each lung. Blood from the capillaries of the lung returns to the heart through the right and left *pulmonary veins,* which empty into the *left auricle.* From the left auricle it passes into the *left ventricle* through an opening guarded by the *mitral (bicuspid) valve.* The left ventricle pumps the blood into the systemic circulation through the *aorta,* in which there are three semilunar valves at the origin. The aorta arches around the dorsal side of the heart on the left side. (Its main branches are described below.) Blood from the systemic circulation returns to the *right auricle,* from which it passes into the right ventricle through an opening guarded by the *tricuspid valve,* thus completing its circulatory pattern (Fig. 15.3).

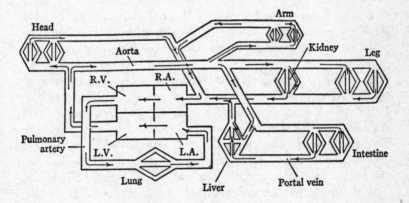

Fig. 15.3. Diagram illustrating the direction of blood flow in man. Applies equally well to all mammals and (by substituting "wing" for "arm") to all birds. The aortic arch swings to the left in mammals, to the right in birds.

Branches from the arch of the aorta are: the *innominate artery*— dividing into *right subclavian* and *right carotid arteries,* the *left carotid artery,* and the *left subclavian artery.* The carotids extend into the head; the subclavians, through the shoulder into the arm. The *descending aorta* gives off branches to muscles, lungs, and abdominal viscera, and divides in the lower portion of the abdominal cavity into the *common iliac arteries,* which carry the blood into the legs.

The systemic venous system consists of: a *portal vein,* carrying blood from the alimentary canal to the liver (a renal portal vein—such as occurs in the frog—is absent) ; the *inferior vena cava,* a large vein draining the lower part of the body and emptying into the *right auricle;* and the *superior vena cava,* formed by the union of the *right* and *left innominate veins,* each of these formed by a *jugular* and a *subclavian vein.* The jugulars drain the head; the subclavians, the arms. Microscopic blood vessels, the *capillaries,* carry the blood directly to the tissues of all parts of the body. They provide the connection between arteries and veins and between the portal system and the inferior vena cava, and, from the point of view of circulatory function, are the most important structures of the circulatory system.

LYMPHATIC SYSTEM. Intercellular spaces, which are filled with material that diffuses from the capillaries, are drained by a set of channels called *lymph vessels* that empty into the subclavian veins near the heart. The lymph vessel on the left side, the *thoracic duct,* drains all the body below the diaphragm as well as the left side above it. The right lymphatic duct, which drains the right side of the body above the diaphragm, is smaller.

**Physiology.** The course of the blood, outlined in the description of the morphology of the heart and blood vessels, is diagrammed in Figure 15.3. Functions of the blood include (1) transport of food material and oxygen to the cells and waste material away from them, (2) regulation of body temperature, (3) transport of hormones, and (4) disease resistance. Blood capillaries in the walls of the intestine absorb end products of carbohydrate and protein digestion. These are conveyed to the liver by the portal vein. Products of fat digestion are absorbed by the *lacteals* (small lymph channels in the villi), through which they are conducted to the veins by way of the thoracic duct. The disease-resisting function is accomplished by the *antigen-antibody reactions* of the blood, a foreign protein (antigen) in the blood stimulating the formation of an antibody. The latter may *precipitate* the foreign protein if the protein is in solution or *agglutinate* the foreign bodies into clumps if they are insoluble. In the latter case, *phagocytosis* (the engulfing of the foreign material by white corpuscles) is increased. A poison (*toxin*) produced by a foreign organism stimulates production of an *antitoxin,* a kind of antibody.

There are sufficient differences among human blood proteins to result in agglutination of erythrocytes if incompatible bloods are mixed during transfusion. Four major types are recognized: *Types O, A, B,* and *AB.* These types are named on the basis of the antigens in their erythrocytes. Type O has both types of agglutinins in its plasma but no antigens in the erythrocytes, hence is the "universal donor." Type AB, which has both antigens in its corpuscles but no agglutinins in its plasma is called the "universal recipient." (There are other differences among human blood types that must also be considered during transfusions.)

Blood clotting depends upon release of *thromboplastin* from the thrombocytes. Thromboplastin, an enzymatically active substance in the plasma, converts *prothrombin,* which is already present in the plasma, into *thrombin.* (Vitamin K is involved in prothrombin synthesis. Calcium ions are required in the production of thrombin.) Thrombin transforms *fibrinogen,* a protein in the plasma, into *fibrin,* the material of the clot. As the fibrous material of the clot contracts it squeezes out a liquid, the *serum.*

The erythrocytes function primarily in transport of oxygen from the lungs to the cells. The oxygen is carried in loose combination with the protein *hemoglobin,* in the red cells. The combination, *oxyhemoglobin,* formed in the capillaries of the lungs, is readily reduced in the cells merely by exposure to the lower oxygen pressure.

## RESPIRATORY SYSTEM

**Morphology.** The air passages in the nose are increased in surface area by *turbinal bones,* over which extends the olfactory membrane. The passages open into the roof of the *pharynx* at the back of the *soft palate.* From the pharynx, the air passage leads into the *larynx* ("voice box") through an opening covered by a flap, the *epiglottis.* The larynx is at the upper end of the *trachea* or windpipe. The latter bifurcates, forming two *bronchi,* one entering each *lung,* where it repeatedly divides into the smaller *bronchial tubes;* these terminate in thin-walled pockets, the *alveoli,* arranged in groups. The pulmonary capillaries are in the walls of the alveoli.

Each lung is covered by a membrane, the *visceral pleura,* which is reflected back at its root to line the cavity in which the lung lies. The outer wall of the cavity is lined with the *parietal pleura.* Each of the two pleural cavities is separate, and neither communicates with any other portion of the coelom.

**Physiology.**

BREATHING. The movement of air into and out of the lungs properly constitutes *breathing,* not "respiration," in spite of popular usage to the contrary. The mechanics of breathing involve movements of the floor and wall of the chest cavity. The floor is formed by the diaphragm; increase in height of the chest cavity (actually the two pleural cavities) results from contraction of this muscle. Increase in transverse diameter results from raising the ribs, this movement produced by contraction of *intercostal muscles.* Because the parietal pleura is attached to the outer wall, increase in size of the chest is accompanied by withdrawal of the parietal pleura from the lungs. Decrease of air pressure in the pleural cavity then decreases air pressure in the lungs, and, if the air passages are open, air is drawn into the lungs.

OXYGEN TRANSPORT. Because alveolar air normally contains more oxygen and less carbon dioxide than does the blood in the pulmonary capillaries, oxygen tends to diffuse into the blood and carbon dioxide from the blood into the alveoli. Once in the blood, most of the oxygen is picked up and carried in loose combination with *hemoglobin,* the reddish pigment of the erythrocytes, the combination being called *oxyhemoglobin.* In the capillaries in the systemic (nonpulmonary) portion of the blood system the concentration gradient is in the opposite direction from that in the lung wall, the oxygen being more concentrated in the blood than in the cells. The oxyhemoglobin is spontaneously reduced, under the lower oxygen pressure, and the oxygen diffuses into the cells. Carbon dioxide, more concentrated in the cells than in the blood, diffuses into the blood and is carried to the lungs.

CELL RESPIRATION. The oxidation that takes place within the individual cells is, properly speaking, cell respiration. This process, essentially the same in all animals and plants, has already been described in Chapter IV.

## EXCRETORY SYSTEM

**Morphology.**

KIDNEYS. The two bean-shaped kidneys, each about four inches long, lie in the posterior wall of the peritoneal cavity. In section, each appears to have three regions, a medial cavity, the *pelvis,* a central portion, the *medulla,* and an outer region, the *cortex.* From each kidney a duct, the *ureter,* conveys urine to the *urinary bladder.* The channel by which urine leaves the bladder is the *urethra* (Fig. 15.4*A*). The urethra also carries reproductive products in the male, and its external opening in the female is adjacent to that of the reproductive channel. The secretory units of the kidneys, the *nephridia,* are microscopic structures in the cortex. Each begins as a double-walled, hollow sac, *Bowman's capsule,* which leads into a long tubule made up, in sequence, of the *proximal convoluted tubule,* the *loop of Henle,* and the *distal convoluted tubule* (Fig. 15.4*B*). This last drains into a *collecting tubule* that carries urine to the pelvis of the kidney.

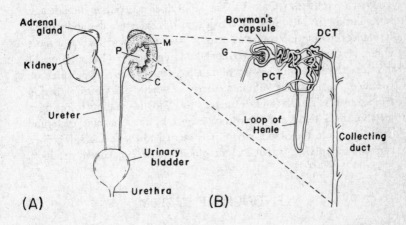

Fig. 15.4. (*A*) The kidneys and associated structures, from the ventral side. The left kidney, in horizontal section, reveals its two layers (*C*, cortex; *M*, medulla) and its cavity (*P*, pelvis) leading into the ureter. (*B*) A nephridium, its orientation in the kidney indicated by the projecting broken lines. Bowman's capsule leads into the proximal convoluted tubule (*PCT*), loop of Henle, distal convoluted tubule (*DCT*), and a collecting duct that empties into the pelvis. The blood capillaries include a knot (*G*, glomerulus) inside Bowman's capsule and a network around the tubules, both more elaborate than here indicated. (These drawings, considerably simplified, are modified from numerous sources.)

OTHER EXCRETORY ORGANS. The lungs, the sweat glands, and also, in the strictest sense, the liver are all organs with excretory functions.

**Physiology.**

KIDNEYS. Urine is formed in the nephridium by a combination of dialysis and active transport. It begins as a watery filtrate of the blood plasma in Bowman's capsule, where it is derived from a net of blood capillaries, the *glomerulus,* enclosed in the capsule. The force producing the dialysis is blood pressure. Most of the water is reabsorbed, and, with it, various compounds in the filtrate that are not waste products— glucose, amino acids, certain salts. Most of this reabsorption takes place in the proximal convoluted tubule. It obviously requires movement of materials from inside the tubule into the bloodstream—against the diffusion gradient. This is accomplished by secretory cells, making use of metabolic energy for active transport. The final product contains high concentrations particularly of nitrogenous wastes, urea being the most abundant of these in human urine.

OTHER EXCRETORY ORGANS. The lungs discharge carbon dioxide and water during the breathing cycle. Sweat glands excrete water and various substances in solution, but this activity is in part incidental to a more important function in man, temperature regulation. (Evaporation of moisture from the skin has a cooling effect, and the rate of perspiration is under control of the autonomic nervous system and varies with skin temperature.) The liver has excretory functions in relation to nitrogen metabolism. Urea is produced in the liver by the ornithine cycle and discharged into the blood, where it is carried to the kidneys. The bile contains bile pigments that include blood protein breakdown products.

# ENDOCRINE SYSTEM

Although not constituting a definite organ system, the *endocrine glands,* which secrete *hormones,* may be summarized together. In all cases their secretory products are picked up and carried in the bloodstream rather than in specialized ducts; hence they are called endocrine glands, meaning glands of internal secretion. The hormones, which produce effects remote from their place of origin, are chemical co-ordinators.

The *pituitary gland,* at the base of the brain, consists of two main divisions, the *anterior* and *posterior lobes.* The former secretes hormones that regulate growth, sexual development, and the activity of other endocrine glands. The posterior lobe secretes hormones that affect smooth muscle activity and the water relations of the circulatory system. The *thyroid gland,* a bilobed structure in front of the trachea, secretes a hormone (*thyroxin*) that regulates basal metabolism. The *parathyroid glands,* four small bodies that may be embedded in the thyroid, secrete a hormone that regulates calcium balance between the bones and the blood. The *thymus,* also in the region of the throat, is necessary for establishing the body's mechanisms of immunity; its importance decreases after early development. In the pancreas occur scattered groups of endocrine cells, the *islets of Langerhans,* which secrete *insulin,* a hormone involved in glycogen storage in the liver—and therefore in the sugar balance in the blood. The *adrenal glands,* which are endocrine, lie anterior to and in contact with the kidneys. They secrete a hormone (*epinephrine* or *adrenalin*) from the *medulla* or central portion and a mixture of hormones (collectively called *cortin*) from the *cortex.* Epinephrine produces the same effects as those accomplished by stimulation of the sympathetic portion of the autonomic nervous system —acceleration of the heart beat, heightened blood pressure, inhibition of secretion of certain glands, and other effects. Cortin is involved in the regulation of metabolism and in reproduction. The parathyroid secretion, insulin, and cortin are essential to life. A few hormones that regulate the production of digestive secretions are formed in the walls of the duodenum. Hormones responsible for the development of secondary sexual characteristics are secreted by cells in the gonads of both sexes, and from the ovary arise hormones involved in the regulation of menstruation, pregnancy, and lactation. (Reproductive hormones are considered in more detail in Chapter XVI.) In addition to the glands of known endocrine function there is at least one other, the *pineal body,* a tiny organ on the dorsal side of the brain, suspected of such function.

## NERVOUS SYSTEM

The nervous system consists of the *central nervous system* (*brain* and *spinal cord*), the *peripheral nervous system* (*cranial* and *spinal nerves*), and the *autonomic nervous system.*

**Morphology.**

CENTRAL NERVOUS SYSTEM. The brain has five major divisions. The *cerebral hemispheres* (derived from the embryonic *telencephalon*) are so large in the human brain that the *diencephalon* and *midbrain* (*mesencephalon*), the next two divisions, are completely concealed by them. The *cerebellum* (*metencephalon*) is, however, relatively large. These four, with the *medulla oblongata* (*myelencephalon*), constitute the five divisions of the brain. The medulla merges into the *spinal cord,* which has two enlarged regions, the *cervical* and *lumbar enlargements.* (Fig. 11.23.)

The brain and spinal cord are surrounded by three membranes or *meninges.* These are the *pia mater,* next to the brain or cord; the *dura mater,* lining the cranium and neural canal of the vertebrae; and the loose *arachnoid membrane* between them.

PERIPHERAL NERVOUS SYSTEM. There are twelve cranial nerves, as in reptiles, birds, and other mammals, with essentially the same distribution. Three are purely sensory: I, *olfactory;* II, *optic;* VIII, *auditory.* Three (III, IV, and VI) serve the eye muscles. The *vagus nerve* (X) extends into the thoracic and peritoneal cavities from the head. The other cranial nerves supply head and neck with sensory and motor fibers. Of these nerves, the *trigeminal* (V) and the *facial* (VII) are quite important. Man has thirty-one spinal nerves: eight *cervical,* twelve *thoracic,* five *lumbar,* five *sacral,* and one *coccygeal.* The spinal nerves join to form *cervical, brachial, lumbar,* and *sacral plexuses* in man. Each spinal nerve has two *roots,* a *dorsal* or *sensory root* and a *ventral* or *motor root.* The nerve branches into three *rami,* a dorsal, a ventral, and a communicating ramus (containing fibers of the autonomic nervous system). (Fig. 15.5.)

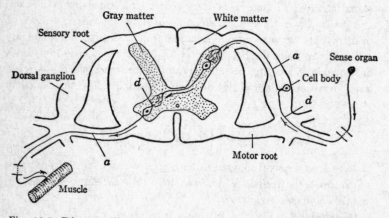

Fig. 15.5. Diagram illustrating relations between spinal cord and spinal nerves. Three neurons, constituting a simple reflex arc, are represented. The direction of the nerve impulse through these neurons is indicated by arrows. Abbreviations: *a,* axon; *d,* dendrite.

AUTONOMIC NERVOUS SYSTEM. The autonomic nervous system is related structurally to the central nervous system but the actions it controls are independent of the will. The system consists of two parts, the *sympathetic system,* arising in the thoracic region, and the *parasympathetic system,* arising in the cranial and sacral regions. Fibers from both systems invade all parts of the body, and the impulses they carry are mutually antagonistic. Impulses in the sympathetic system accelerate the heart beat, slow peristalsis, and produce dilation of the iris of the eye. Impulses in the parasympathetic system slow the heart beat, speed peristalsis, and produce constriction of the iris.

**Physiology.** The principal function of the nervous system is the conduction of *nerve impulses* from one part of the body to another, thus bringing about coordinated responses. The cells involved are the highly specialized *neurons,* which are associated together in nerve tracts and reflex arcs. A simple reflex arc (Fig. 15.5) involves at least a *receptor* (sense organ), an *afferent* (sensory) *neuron,* an *efferent* (motor) *neuron,* and an *effector* (motor organ). From a few to many *associating neurons* are nearly always present between afferent and efferent neurons. The connections between neurons (synapses) are not intimate. Most of the delay in the passage of a nerve impulse through a reflex arc is due to the number of synapses over which it must pass rather than the total distance it travels. Transmission across the synapse is mediated chiefly by secretions called *neurohumors.* The neurohumor released by the vagus (inhibitor) nerve endings in heart muscle is *acetylcholine.* (The same substance is released at nerve-muscle connections in skeletal muscle.) The neurohumor released where accelerator nerve fibers of the sympathetic system end in heart muscle is, however, *sympathin* (similar to or the same as the hormone epinephrine). The physiology of the transmission of nerve impulses has been described in Chapter XII.

## SENSORY SYSTEM

Among the sense organs of man, the eye and the ear are especially complex in their specializations. The former is adapted for vision; the latter, for two sensory functions, hearing and equilibration.

**Eye.** (Fig. 15.6.) As previously stated (p. 194) the anatomy of the eye is best understood in analogy with a camera. The human eye has a *lens* and, in addition, two other refractive media, which are liquid. These are the *aqueous humor,* in front of the lens, and the *vitreous humor,* behind it. The *iris,* in front of the lens, regulates the amount of light impinging on the *retina,* the structure that lines the back of the eye. The retina contains two general types of light-sensitive cells: the *rods,* sensitive to light intensity, and the *cones,* which are color sensitive. Rods

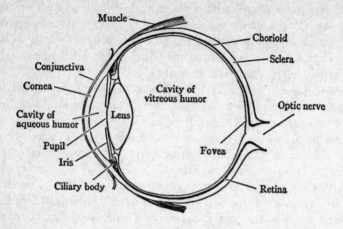

Fig. 15.6. Vertical section through the human eye. The relations, though not the proportions, are typical of vertebrates in general. (Modified from various authors.)

contain *rhodopsin* (*visual purple*), while cones contain pigments sensitive to specific wavelengths (colors) of light. The brain, of course, gives man the final interpretation of visual stimuli, after impulses set up in the retina are conducted to the brain over the optic nerve.

**Ear.** (Fig. 15.7.) Two different sense organs are contained in the ear, the organ for the sense of equilibrium and the organ of hearing. The former consists of three *semicircular canals* at right angles to each other (two vertical and one horizontal), containing nerve endings stimulated by the flow of liquid in the canals. The organ of hearing involves three regions, the *outer ear, middle ear,* and *inner ear.* The outer ear includes the *pinna,* or ear flap, and the *external auditory meatus* or canal, terminating at its inner end with the *tympanic membrane.* Movement of the tympanic membrane is conveyed as vibrations across the middle ear through a chain of three small bones or auditory ossicles, the *malleus* ("hammer"), *incus* ("anvil"), and *stapes* ("stirrup"). These vibrations are transferred to liquid in the inner ear, which contains the real organ of hearing, the *organ of Corti.* Ability to distinguish between sounds of different pitch involves, according to a widely held theory, response by sympathetic vibration to different frequencies at different levels along the organ of Corti. The interpretation of the sensation of sound, as of sight, rests, of course, in the brain.

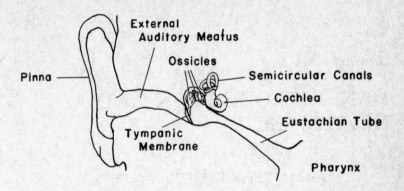

Fig. 15.7. Diagram illustrating the relations of external ear (pinna and external auditory meatus, bounded internally by the tympanic membrane); middle ear (cavity containing the auditory ossicles, communicating with the pharynx through the Eustachian tube); and inner ear (semicircular canals and cochlea). The ossicles are oriented—unnaturally—so that all three show in this plane: malleus, incus, and stapes, in order from left to right. The malleus is attached to the tympanic membrane. The stapes fits into the partition between middle and inner ears. (Simplified, from numerous sources.)

**Other Sense Organs.** In general, other sense organs involve relatively simple terminal nerve elements that do not occur in complexly differentiated organs. The senses of taste, smell, pain, pressure, temperature, and others are of this general type.

*Chapter XVI*

# MAN: REPRODUCTION AND DEVELOPMENT

## MALE REPRODUCTIVE ORGANS

The gonads of the male, the *testes* (Fig. 16.1), are contained in a saclike extension of the coelom, the *scrotum*. Each testis contains many tubules in which spermatozoa are formed and from which numerous *vasa efferentia* drain into the single *vas deferens* from each of the two testes. Each vas deferens or sperm duct extends into the peritoneal cavity, where it meets the urethra and the sperm duct from the opposite side, after receiving the secretions of a *seminal vesicle*. The ducts in this region pass through the *prostate gland,* which contributes to the secretion known as the *seminal fluid*. The urethra, after receiving the two sperm ducts, traverses the *penis*. This intromittent organ consists of three columns of spongy, *erectile tissue* that becomes turgid as a result of the concentration of blood in its vascular spaces. The urethra is the common channel for urine and seminal fluid.

## FEMALE REPRODUCTIVE ORGANS

The two gonads of the female, the *ovaries* (Fig. 16.2), lie in the lower part of the abdominal cavity near the open ends of the *oviducts* or *Fallopian tubes*. The two tubes join to form the *uterus,* which communicates with the external opening, the *urogenital sinus* or *vestibule,* through the *vagina*. The opening of the urogenital sinus is bounded laterally by two pairs of folds, the inner *labia minora* and the outer *labia majora*. The *clitoris,* homologous with the penis of the male, lies at the anterior margin of the urogenital sinus, and the urethra opens into the sinus just posterior to it.

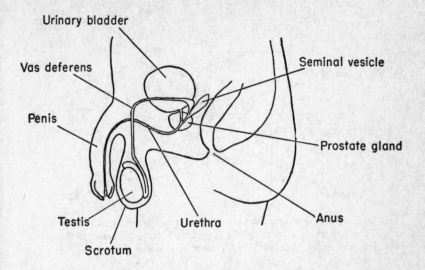

Fig. 16.1. Diagram of the human male reproductive organs, in section, from the left side. (Reprinted by permission from *General Biology* by Gordon Alexander, published by Thomas Y. Crowell Company.)

## THE FEMALE SEX CYCLE

The developing *ovum* is contained in a *follicle* in the ovary, and this follicle grows as the egg cell grows. When fully formed, the follicle ruptures, releasing the ovum. This process is *ovulation*. The ovum is drawn into the Fallopian tube, where fertilization (conception) may occur if motile sperm cells are present. (The sperm cells move through the uterus from the vagina, and up the Fallopian tube, following copulation, and fertilization normally occurs at the upper end of the Fallopian tube. Sperm cells live in the tubes no longer than one to three days so copulation must take place near the time of ovulation for fertilization to occur.) If fertilization does not occur, the ovum is carried down the tube, through the uterus, and it leaves the body through the vagina. In that case, the *corpus luteum*, formed at the surface of the ovary from the ruptured follicle, gradually decreases in size; and in about two weeks after ovulation the lining of the uterus goes through a process of rapid disintegration, the process of *menstruation*. Ovulation

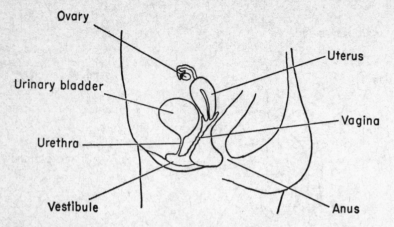

Fig. 16.2. Diagram of the human female reproductive organs, in section, from the left side. (Reprinted by permission from *General Biology* by Gordon Alexander, published by Thomas Y. Crowell Company.)

and menstruation, two events in the *menstrual cycle* of approximately twenty-eight days, are normal aspects of the sex cycle of a woman between *puberty* and *menopause,* except when interrupted by *pregnancy.*

These events in the female sex cycle are related to a corresponding cycle of hormone activity, initiated by the pituitary gland. Hormones involved include the *follicle-stimulating hormone* (FSH) and the *luteinizing hormone* (LH). Following menstruation a new follicle begins to grow, and as it grows it produces the hormone *estrogen.* This hormone initiates the repair and thickening of the uterine wall. Following ovulation, which occurs on the average 14 days after the onset of the preceding menstrual period, the follicle constricts and becomes converted into a corpus luteum, which secretes the hormone *progesterone.* This continues the influence on the uterine wall begun by estrogen. If fertilization does not occur the corpus luteum disintegrates and ceases to produce progesterone; it is in the interval before estrogen from a newly developing follicle takes over that menstruation occurs.

## EMBRYONIC DEVELOPMENT

If fertilization (conception) occurs the menstrual cycle is arrested. Development of the zygote begins in the upper end of the oviduct, and it continues as the embryo passes down the tube into the uterus. Cleavage is holoblastic, but the early embryo is a somewhat modified blastula (*blastocyst*). The blastocyst is a hollow ball of cells, as is a typical blastula, but with an *inner cell mass* suspended within from one side. The outer layer, the *trophoderm,* is eventually involved in forming part of the placenta, while the inner cell mass becomes the embryo and embryonic membranes associated with it.

Formation of the blastocyst has occurred by the time *implantation* of the embryo in the wall of the uterus has taken place, ordinarily about nine days after fertilization. (The intrauterine devices, IUD, used for contraception probably prevent implantation, but their method of functioning is unknown.) If implantation occurs the corpus luteum persists and grows, and it continues to secrete progesterone. This hormone helps maintain the embryo in the uterus and, at the same time, prevents further ovulation during pregnancy. The developing placenta, whose formation begins at the site of implantation, also secretes progesterone. (The contraceptive pills containing similar compounds function by preventing or regulating ovulation.)

At about the time of implantation the inner cell mass undergoes gastrulation by dividing into two groups of cells: those destined to be ectoderm, which are attached to the inside of the trophoderm, and those destined to be endoderm, on the side toward the cavity of the blastocyst. Each cell mass now develops a cavity within it, that in the ectodermal mass becoming the amniotic cavity, that in the endodermal mass becoming the yolk sac. The embryo differentiates in the region where the two cell masses meet, and the mesoderm develops between these two layers.

As the mesoderm grows out from the region that is to become the posterior end of the embryo it meets the trophoderm. This structure of mesoderm and ectoderm (for the trophoderm is considered ectoderm), the *body stalk,* is the forerunner of the *umbilical cord,* which connects embryo and placenta. The blood vessels that connect embryo and placenta develop in the mesoderm of the body stalk. These are the *umbilical vein,* carrying blood from the placenta, and the two *umbilical arteries,* carrying blood to the placenta. The placenta is of double origin: embryonic tissue of mesoderm and ectoderm, and maternal tissue from the lining of the uterus. The relation between embryonic and maternal blood vessels is intimate, but there is no direct connection (and there is no direct connection of the nervous systems or any other organ systems between embryo and mother). The yolk sac is vestigial, the human egg having practically no yolk; the nutrition of the developing embryo (*fetus*) is derived by diffusion from the maternal bloodstream. The fetus is enveloped in the amniotic cavity, which is filled with liquid and which expands with growth of the fetus.

At the end of the period of pregnancy, about ten lunar months after fertilization, birth or *parturition* occurs. The time of parturition is calculated in advance as 280 days after the beginning of the last menstruation (since menstruation ceases during pregnancy). Shortly after parturition, *lactation* or milk secretion begins, initiated by the hormone *prolactin,* produced by the pituitary gland. Lactation continues through stimulation caused by drainage of the ducts of the *mammary glands.*

*Part Four*

# GENERAL  PRINCIPLES

# HEREDITY

*Heredity* is the transmission of traits, physical or mental, from parents to offspring. The scientific study of this transmission is *genetics;* it may follow either statistical or experimental methods.

## THE CONTINUITY OF THE GERM PLASM

In 1883, August Weismann suggested that all hereditary change must originate in the germ plasm, inasmuch as *the germ plasm constitutes the only organic continuity from one generation to the next.* This concept of the continuity of the germ plasm is the basic principle in all studies of biological inheritance. (See Fig. 18.3.)

## MENDELISM

**Gregor Mendel.** Working on garden peas, studying the effects of crossing peas of contrasting characteristics and their descendants, Gregor Mendel discovered in the early 1860's the principles now known as Mendel's laws. Their true scientific value was not appreciated until 1900, after Mendel's death.

**Mendel's Experiments.** These involved seven pairs of contrasting characters of the garden pea. Mendel discovered that a cross of plants that each bred true for a contrasting form of a character (e.g., smooth seeds vs. wrinkled seeds) produced progeny that all resembled one parent only. When the flowers of these (the $F_1$ generation) were permitted to be self-fertilizing, however, both of the original parental forms were present in their progeny. These kinds were present in numbers approximating a 3 : 1 ratio—three like the $F_1$ to one like the parental form that did not appear in the $F_1$ generation. The latter

Mendel called *recessive;* the former, *dominant.* Seeds from plants showing the recessive trait produced plants all having the recessive trait; but seeds from a plant showing the dominant trait but having the recessive trait in their ancestry developed into both types of plants.

**Interpretation of Mendel's Experiments.** The importance of the gametes was realized by Mendel, who saw that to explain the ratios he obtained he had to assume that a given gamete could contain a determiner for one character but not its opposite. If the determiner for the dominant trait was derived from both parents or *from either one,* the offspring would show the dominant trait; the recessive trait appeared only when the gametes from *both parents* contained its determiner. Each trait appeared as if determined by a single factor in each gamete —the gametes being themselves "pure" for one of the two contrasting characters. This explanation coincided also with Mendel's observation that, regardless of the number of contrasting characters studied at one time, the hereditary behavior of each one was independent of the others.

**Mendel's Laws.** Aside from the idea of dominance, which is not universally true, Mendel's chief contributions were the concepts of (1) *segregation* and (2) *independent assortment.* A corollary of the former is the concept of the *purity of the gametes.* The genetic basis for these two laws of Mendelism is the process of meiosis (see Chap. V and below, in this chapter). Independent assortment occurs, however, only when the determiners are in different pairs of chromosomes. It does not occur when the determiners show linkage (see below).

**Terminology of Mendelian Inheritance.** Although much of this terminology has developed since Mendel's time it is applied to the same types of observations he made. Unit characters may be determined by single *factors* or *determiners,* now called *genes.* Alternative characters are *allelomorphic;* their determiners cannot occur in the same gamete. A zygote containing two determiners for the same character is *homozygous;* one containing the genes for alternative characters is *heterozygous.* All organisms that look alike with reference to alternative characters belong to the same *phenotype;* all organisms of the same genetic behavior with reference to alternative characters belong to the same *genotype.* (To illustrate: heterozygous dominants and homozygous dominants are of the same phenotype but different genotypes.)

Mendel referred to the first individuals he crossed as the *parental generation* (P) and to their progeny as the *first filial generation* ($F_1$). Crossing $F_1$ individuals results in the production of the $F_2$ generation. In order to determine the genotype of a known phenotype a *test cross* is used. The individual in question is crossed to a homozygous recessive individual, in which the gametes are therefore known, and the resulting ratio of $F_1$ individuals reflects the gametes formed and therefore the genotype of the organism in question. This test cross is also called a *back cross* because Mendel crossed individuals in question back on the original pure breeding stock. A cross in which we consider only one pair of contrasting characters is called a *monohybrid cross,* while one in which we consider two pairs of contrasting characters simultaneously is called a *dihybrid cross.* (Note: the term hybrid does not here refer to a cross between different species, as is usually the case, but to a cross between genetically different individuals of the same species.)

**Examples of Mendelian Inheritance.** While it is not true that all inheritance of unit characters is as simple as was the case for those Mendel studied, yet it is necessary to understand such simple situations in order to comprehend more complex ones.

EXAMPLES FROM THE GARDEN PEA. Three contrasting characters studied by Mendel had to do with length of stem, color of seed, and character of seed coat. Mendel found that the tall stem condition was dominant over dwarf, that yellow seed color was dominant over green, and that a smooth seed coat was dominant over a wrinkled one. The following outlines illustrate the results in these crosses:

(P = parental generation      $F_1$ = first filial generation)

| | |
|---|---|
| P | Tall × Dwarf |
| $F_1$ | All Tall |

Various combinations failed to modify the principle of dominance:

| | |
|---|---|
| P | Tall Yellow Smooth × Dwarf Green Wrinkled |
| $F_1$ | All Tall Yellow Smooth |

It is obvious, however, that tall plants from dissimilar parents have really developed from gametes containing the determiners for *both* tall and dwarf. Hence:

(T = gene for tall                t = gene for dwarf
  Y = gene for yellow             y = gene for green
    S = gene for smooth         s = gene for wrinkled
          and G = gametes)

| | | |
|---|---|---|
| P | TT (homozygous tall) × | tt (homozygous dwarf) |
| G | All T | All t |
| $F_1$ | All Tt (heterozygous tall) | |

P                      TTYYSS × ttyyss
G           All TYS                   All tys
$F_1$      All TtYySs (heterozygous tall yellow smooth)

(Note: Following the principle of the purity of the gametes, a gamete cannot contain two "t's," two "y's," or two "s's"; it will always contain one representative of each group of allelomorphs.)

Mendel found that self-fertilized flowers of heterozygous tall (Tt) plants produced plants both tall and dwarf in approximately a 3 : 1 ratio. The letter method of representation shows how, by chance distribution of genes in gametes and the subsequent chance fertilization of any egg nucleus by any pollen-tube sperm nucleus, this ratio is achieved. In crosses in which one studies two or three pairs of contrasting characters the ratios are different, but all more complicated ratios have definite arithmetical relationship with those found in simple one-factor crosses:

P                     Tt × Tt
G          ½ T, ½t, from each parent
$F_1$        ¼TT    ½Tt    ¼tt

or, the genotypic ratio is:
1 homozygous tall : 2 heterozygous tall : 1 homozygous dwarf; and the phenotypic ratio is 3 tall : 1 dwarf.

(Note: The female gamete represented by T may be fertilized by the male gamete T or t; the chances are equal. The same is true for female gamete t. If one flips two coins simultaneously a great many times, recording results each time, the expected ratio of results is, as here: 25 percent two heads : 50 percent one heads, one tails : 25 percent two tails. See Fig. 17.1.)

P                 Tt × tt
G          ½T, ½t     All t
$F_1$      ½Tt ½tt    (or 1Tt : 1tt)

P           Tt Yy    ×    Ttyy
G    ¼ each: TY, Ty, tY, ty    ½ each: Ty, ty
$F_1$   1TTYy : 1TTyy : 2TtYy : 2Ttyy : 1ttYy : 1ttyy

(The above is the genotypic ratio; the phenotypic ratio is 3 tall yellow : 3 tall green : 1 dwarf yellow : 1 dwarf green.)

EXAMPLES FROM GUINEA PIGS. Certain contrasting characters in the guinea pig behave in inheritance much as the ones given from the garden pea. Black coat color is dominant over white; in the same way, a rough coat shows dominance over a smooth one; these behave as unit characters.

$$(B = \text{gene for black} \qquad b = \text{gene for white}$$
$$R = \text{gene for rough} \qquad r = \text{gene for smooth})$$

P        BbRR      $\times$      bbRr

G       ½BR, ½bR          ½bR, ½br

$F_1$    Genotypes: 1BbRR : 1BbRr : 1bbRR : 1bbRr

       Phenotypes: 2 black rough : 2 white rough

## Problems in Mendelian Inheritance.

METHODS OF SOLVING MENDELIAN PROBLEMS. The two common methods of solving Mendelian problems are the Punnett-square method and the algebraic method.

*The Punnett-square or "Checkerboard" Method* (including all pictorial or graphical methods). This method is too unwieldy for problems that cannot be solved by rapid inspection. It involves finding all possible combinations of all types of gametes from one parent with all types of gametes from the other parent, duplicate zygotes to be sorted out and grouped together after all combinations are made. The method is illustrated in Figure 17.1. It illustrates the chance combinations of gametes very well but is too slow for practical use.

*The Algebraic Method.* Inspection of the results of a cross between two heterozygous parents suggests the fundamental relationship between Mendelian crosses and the products of algebraic quantities. To illustrate: Tt becomes $T + t$ after segregation in the gametes. Tt $\times$ Tt becomes, therefore, $(T+t)(T+t)$ or $(T+t)^2$. Just as $(a+b)^2 = a^2 + 2ab + b^2$, so $(T+t)^2 = T^2 + 2Tt + t^2$ (1TT : 2Tt : 1tt); or, in a more complicated cross: Ttyy $\times$ TtYy is the same as $[(T+t)(y)][(T+t)(Y+y)]$. Multiplying out we obtain: TTYy + TTyy + 2TtYy + 2Ttyy + ttYy + ttyy, which is the expected genotypic ratio in the above cross. Now, there are only three possible results in a simple cross involving but one pair of characters: (1) the offspring are all alike if both parents are homozygous; (2) half the offspring are like one parent and half like the other if one parent is homozygous and the other heterozygous; and (3) there is one homozygous dominant to two heterozygous dominants to one homozygous recessive if both parents are heterozygous. The ratios to be expected in more complex crosses are all combinations of these three conditions, and may be analyzed by considering each pair of characters alone, in order. This does not, of course, require any laborious

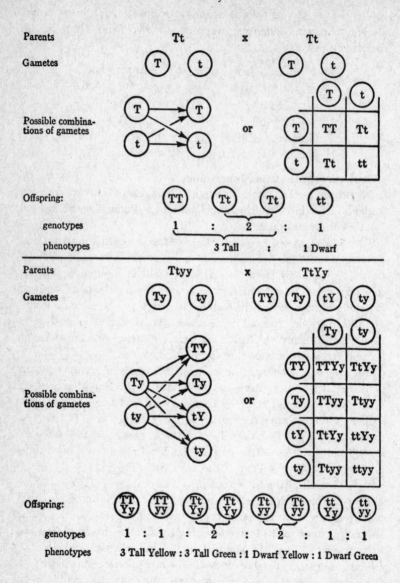

Offspring: genotypes 1 : 2 : 1

phenotypes 3 Tall : 1 Dwarf

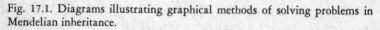

Fig. 17.1. Diagrams illustrating graphical methods of solving problems in Mendelian inheritance.

multiplication. One simply keeps in mind the three possible results in single-factor crosses and applies these ratios to other situations. Applying this method to the above problem (Ttyy × TtYy), we should analyze the inheritance of the tall-dwarf condition first, then the yellow-green. In a Tt × Tt cross we have two heterozygotes, yielding the ratio: 1TT : 2Tt : 1tt. In the other characters, we have one heterozygous and one homozygous parent, yy × Yy, yielding 1yy : 1Yy. We then combine the ratios, as in the following scheme in which all the factors in one ratio are multiplied by all the factors in the other:

$$
1\ TT\begin{cases} 1\ yy \\ 1\ Yy \end{cases} \qquad \begin{matrix} 1\ TTyy \\ 1\ TTYy \end{matrix}
$$

$$
2\ Tt\begin{cases} 1\ yy \\ 1\ Yy \end{cases} \qquad \begin{matrix} 2\ Ttyy \\ 2\ TtYy \end{matrix}
$$

$$
1\ tt\begin{cases} 1\ yy \\ 1\ Yy \end{cases} \qquad \begin{matrix} 1\ ttyy \\ 1\ ttYy \end{matrix}
$$

The last column represents the expected genotypic ratio. Any number of characters can be handled similarly, for example:

AabbCCdd × AaBbccDd

$$
1\ AA\begin{cases} 1\ bb\quad 1\ Cc \begin{cases} 1\ dd & 1\ AAbbCcdd \\ 1\ Dd & 1\ AAbbCcDd \end{cases} \\ 1\ Bb\quad 1\ Cc \begin{cases} 1\ dd & 1\ AABbCcdd \\ 1\ Dd & 1\ AABbCcDd \end{cases} \end{cases}
$$

$$
2\ Aa\begin{cases} 1\ bb\quad 1\ Cc \begin{cases} 1\ dd & 2\ AabbCcdd \\ 1\ Dd & 2\ AabbCcDd \end{cases} \\ 1\ Bb\quad 1\ Cc \begin{cases} 1\ dd & 2\ AaBbCcdd \\ 1\ Dd & 2\ AaBbCcDd \end{cases} \end{cases}
$$

$$
1\ aa\begin{cases} 1\ bb\quad 1\ Cc \begin{cases} 1\ dd & 1\ aabbCcdd \\ 1\ Dd & 1\ aabbCcDd \end{cases} \\ 1\ Bb\quad 1\ Cc \begin{cases} 1\ dd & 1\ aaBbCcdd \\ 1\ Dd & 1\ aaBbCcDd \end{cases} \end{cases}
$$

The phenotypes, usually fewer in number than genotypes, may be found by this method, simply by combining phenotypic rather than genotypic ratios from single-factor crosses.

SAMPLE PROBLEMS IN MENDELIAN INHERITANCE.

a. *Problem:* What is the expected ratio of genotypes and phenotypes in a cross between a homozygous dwarf yellow plant (pea) and one heterozygous tall and homozygous green?

*Answer:* Genotypes: 1ttYy : 1TtYy

Phenotypes: 1 dwarf yellow : 1 tall yellow

b. *Problem:* In a litter of seven guinea pigs, one has black rough hair, three have black smooth hair, two have white rough hair, and one has white smooth hair. The mother has white smooth hair. Describe the father, and give genotypes of both parents and all offspring.

*Suggestions for solution:* The female parent, showing both recessive characters, is, of necessity, a double homozygote. The nearest Mendelian ratio which fits the observation of the litter is 1 : 1 : 1 : 1. This means two 1 : 1 ratios combined; each such results from a heterozygous × homozygous cross. The same type of litter could also result from: Bbrr × bbRr.

*Answer:* The father had black rough hair, being heterozygous for both characters. The mother is homozygous white and smooth. All black and all rough offspring are heterozygous; all white and all smooth offspring are homozygous.

## SEX DETERMINATION

The first clue to the structural basis for Mendelian inheritance came with the discovery that the chromosome constitution of cells differs between the two sexes. Most chromosome pairs are alike in both sexes, but one pair shows consistent differences. In most animals the members of this pair are alike in the female (the two X chromosomes) whereas in the male only one X chromosome is present, accompanied by a dissimilar one (the Y chromosome) or not accompanied at all. The male may be represented by the letters XY (or XO), the female by XX. In segregation, half the gametes of the male contain the X chromosome, the other half the Y or no homologue of the X; all the gametes of the female contain the X chromosome. It is really, therefore, a cross of a homozygous with a heterozygous individual, giving the 1 : 1 expected ratio—approximately the observed ratio of the sexes. Other factors may, of course, change the ratio in mature individuals; this is merely the ratio of zygotes at fertilization. To illustrate:

P $\quad$ XX (female) $\quad$ × $\quad$ XY (male)

G $\quad$ All X $\quad$ $\frac{1}{2}$X, $\frac{1}{2}$Y

$F_1$ $\quad$ $\frac{1}{2}$XX (female) : $\frac{1}{2}$XY (male)

The opposite condition, in which the female is heterozygous, occurs in birds and lepidoptera (moths and butterflies):

| P | XY (female) | × | XX (male) |
|---|---|---|---|
| G | $\frac{1}{2}$X, $\frac{1}{2}$Y | | All X |
| $F_1$ | $\frac{1}{2}$XY (female) : $\frac{1}{2}$XX (male) | | |

In the first type, the gamete from the male determines the sex; in the second type, that from the female. The ratio is not affected. (Various special cases and exceptions occur; i.e., sex-reversal in fishes, amphibians, and birds; and so-called intersexes and gynandromorphs.)

## THE ROLE OF GENES

It was but a step from the discovery of the chromosome basis of sex determination to the realization that pairs of determiners must exist, similarly, in homologous pairs of chromosomes.

**The Gene Theory.** The theory that the determiners of hereditary traits, which are called *genes,* are located in the chromosomes is now solidly based on a large number of statistical data—the most significant studies having been carried out by T. H. Morgan and his associates on the common fruit fly of the genus *Drosophila* (Fig. 17.2). This is still one of the most widely used organisms in experimental genetics. Studies with *Drosophila* have demonstrated that genes for alternative traits (e.g., red eye and brown eye in *Drosophila* or tallness and dwarfness in peas), which are called *alleles,* do not exist in the same chromosome because their positions on the chromosome are identical. They occupy the same *locus* in homologous chromosomes; they therefore cannot be present in the same gamete. Segregation of homologous chromosomes during meiosis (Figs. 5.3 and 13.1) separates homologous chromosomes from each other. Thus, though there are many loci on a single chromosome, only one of a series of alleles can be present in a single gamete. After fertilization, when homologous pairs of chromosomes appear again, two genes of the same kind in the zygote give us the homozygous condition while two alternative genes (alleles) give us the heterozygote.

**Multiple Alleles.** A gene may be represented by a series of alternative alleles rather than by just two possibilities. Thus, though any one chromosome contains only one gene at a particular locus, in some cases there are three or more possible alleles. An example of *multiple alleles*

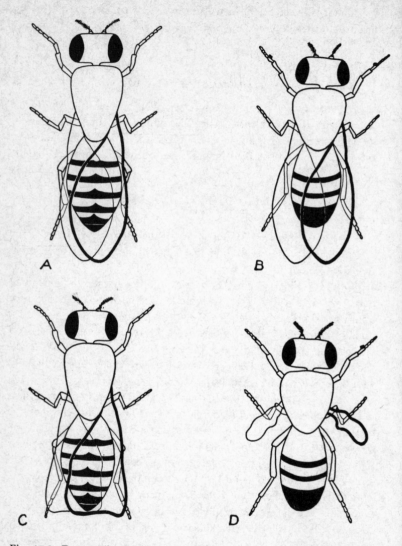

Fig. 17.2. Drosophila. The normal, wild type (*A*, female; *B*, male) has red eyes, long wings, and other features subject to variation through mutation. The sexes are distinguishable by the shape and size of the abdomen and by the presence of the sex comb on the front legs of the male. The mutants shown are (*C*) a female with the wing trait Dumpy, and (*D*) a male with the wing trait Vestigial. (Reprinted by permission from *Laboratory Directions for General Biology* by Gordon Alexander et al., published by Thomas Y. Crowell Company.)

is provided in the inheritance of human AB blood types (see p. 223). These are determined by three alleles, one that results in A type protein production, one that results in B type protein, and one that results in neither type of protein. The various diploid combinations of these alleles result in the different blood types. An unusually large number of alleles occurs in *Drosophila,* where eye color may involve as many as twelve.

**Interaction of Genes.** A particular character may be the resultant of the interaction of two or more genes rather than the result of one alone. These genes may interact to produce a blending, as in what is called *incomplete dominance,* or they may be *complemental,* i.e., two or more genes may be necessary before a given trait becomes evident at all. In addition to these, *lethal genes* are known, their presence in the homozygous condition preventing development of the organism.

**Linkage.** Not all unit characters are independently assorted. The characters Mendel studied in peas were independently inherited, but this is not the general pattern. Certain traits tend always to accompany certain other traits. All these characters that occur together are said to be *linked;* their genes apparently occur in the same chromosome pair. As evidence for this, it has been found that the number of *linkage groups* coincides with the number of chromosome pairs and that one group is typically linked with sex chromosomes. (In *Drosophila melanogaster,* the most widely used species of *Drosophila,* there are four linkage groups. In peas there are seven.) Linkage groups carried in the sex chromosomes are called *sex linked.* The other linkage groups are *autosomal* linkage groups.

**Crossing Over.** During synapsis, in meiosis, the chromosomes of an homologous pair are in intimate contact. During this contact allelomorphic genes may be translocated—genes for alternative characters in some way changing places from one chromosome to its homologue. One explanation given involves the possible twisting of the chromosomes about each other during synapsis. The end result is that certain characters that have been inherited together (linked) may become occasionally separated. One may assume that the further apart linked genes occur on the chromosomes the greater is the possibility of this *crossing over.* Based on that theory, "chromosome maps" have been devised showing the probable position on each chromosome of the genes it is known to bear. The studies of Painter, Bridges, and others on chromosome morphology have yielded a structural as well as theoretical basis for this mapping.

**The Chemical Nature of the Gene.** Before its chemical nature was known the role of the gene in determining the synthesis of an enzyme had been recognized—and we had the "one gene-one enzyme theory" of gene effect. Now, as pointed out in some detail in Chapter V, the gene is known to be a segment of a DNA molecule—though we still do not know how these molecules are oriented in the chromosome. The gene consists of a series of *codons* that, through messenger RNA, controls the synthesis of a polypeptide chain in the cytoplasm. The information for the synthesis is carried in 3-nucleotide groups, "three-letter words," each codon providing the code for a particular amino acid. These codons constitute what is referred to as the *genetic code,* the use of this expression implying that protein synthesis, which is what is controlled by the genetic code, is the means by which genetic traits are manifest. The series of codons determining a particular polypeptide chain is a *cistron;* this may correspond to what we have called a gene in the past. A particular enzyme synthesis may, however, involve other DNA entities, called *operator genes* and *regulator genes,* as well as additional cistrons. Details of the interactions of these factors constitute much of the present research in biochemical genetics. The major problem in interpreting the action of genes—viz., how the same set of genes in all cells can produce differences in different parts of an organism, as well as different organisms—is still with us. This, of course, involves research in cellular biology as well as in genetics and developmental biology.

# HUMAN HEREDITY

**Galton's Laws.** Francis Galton, in the 1880's, enunciated two principles from his statistical studies of human inheritance: (1) the *law of ancestral inheritance,* viz., each parent contributes on the average about one-fourth of the inherited traits, each grandparent one-sixteenth, each great-grandparent one sixty-fourth, etc.; (2) the *law of filial regression,* viz., the offspring of unusual parents tend to be more nearly average than their parents. While in general valid, these laws are of interest to us only historically. (It should be noted that Galton and Karl Pearson were the founders of *biometrics,* the application of statistical analysis to biology.)

**Mendelian Inheritance.** Man has 23 pairs of chromosomes. Twenty-two of these are autosomes; the sex chromosomes are typically XX in the female and XY in the male. Various estimates of the number of gene loci in man have been made, a reasonable estimate being around 25,000. These should be in 23 linkage groups, of which one is sex-linked. Studies of human inheritance are difficult for several reasons, one of the most obvious being the length of a human generation. Nevertheless, a good deal is already known about human inheritance. It involves sex-linkage, multiple factors, and various other complications, and data suggest simple Mendelian inheritance for some traits. For example, brown eye color is generally dominant over blue, curly hair over straight, and dark hair over light. Certain abnormalities are inherited as dominants over the normal condition, e.g., excessively short digits, extra digits, and hereditary cataract. Others may be recessive to the normal condition, e.g., albinism and deaf-mutism. Certain abnormalities appear as sex-linked recessives, appearing in males with a single factor but in females only when there are two, e.g., hemophilia and red-green color blindness. In one case, an extra chromosome is associated with human abnormality: "mongolism" is correlated with three No. 21 chromosomes instead of the normal two.

## THE ORIGIN OF VARIATIONS

The structural basis of heredity explains the persistence of traits, but not their modifications. Changes in the genetic constitution of an organism may be due to new combinations, to chromosome mutations, or to gene mutations.

**New Combinations.** The crossing of two heterozygous individuals may produce new combinations of characters not previously present; the combinations are new, not their individual elements. Crossing over also results in new genetic combinations.

**Chromosome Mutations.** Following synapsis in gametogenesis, the pairs of chromosomes do not always separate normally. In rare cases all or most may go into one germ cell, leaving none or few in the other. Following fertilization, the chromosome number is abnormal; e.g., it may be triploid if one gamete contained the reduced number but the other was diploid, or it may be tetraploid if both gametes were diploid. Such organisms may show traits entirely different from those of either parent, yet heritable. Similarly, dislocation and resultant *deletions, inversions, translocations,* and *duplications* of parts of chromosomes may result in heritable variations.

**Gene Mutations.** Traits may disappear or entirely new ones appear in a given line of descent. If these changes are inherited, they may be due to the loss, gain, or modification of genes in particular chromosomes, *gene mutations,* in other words. Modification of a gene can be due to deletion, addition, or substitution of one or more nitrogen bases in the nucleotides involved. A change in a single codon can be reflected as a gene mutation. These mutations appear "spontaneously," which is to say we do not know their cause; but their frequency of appearance may be increased by X rays or other forms of high-energy radiation, by high temperature, and by certain chemicals.

## EXTRACHROMOSOMAL INHERITANCE

As most observations indicate, and as the gene theory of inheritance implies, inherited traits are derived equally from the two parents. Not all inherited traits are transmitted by the chromosomes, however. As pointed out in Chapter XIII, the nuclear contributions to the zygote are equal but the cytoplasmic contributions of the two parents are not. The egg cell is usually much larger than the sperm cell, for one thing. In any case, some of its cytoplasmic constituents are preserved in the zygote—without equivalent structures from the male gamete. In plants, characteristics of plastids in crosses between parents of different plastid types may be determined by the maternal parent. And in animal development, the pattern of differentiation in the early embryo may be largely determined by cytoplasmic constituents of the egg cell. Both of these cases are examples of what is sometimes called "maternal inheritance," a kind of extrachromosomal inheritance.

## INHERITANCE OF ACQUIRED CHARACTERISTICS

In spite of the obvious adaptation of organisms for their environments, there is no positive evidence for the origin of species through direct action of the environment. The production of heritable changes (mutations) by X rays has involved direct effects on the germ cells. Somatic or body changes are not inherited—apparently because there is no mechanism by which they affect the germ cells.

# Chapter XVIII

# ORGANIC EVOLUTION

*Organic evolution* is the progressive development of animals and plants from ancestors of different forms and functions. It is usually a very slow process, being measured in geological time. The general term *evolution* is applied to any increase in complexity through time—as the evolution of the solar system, the evolution of human society, etc.

## EVIDENCE FOR ORGANIC EVOLUTION

Most evidence for organic evolution is indirect, its validity being supported by many different lines of evidence. This evidence is such that the only scientific explanation for much that we observe in nature is organic evolution. One of the best evidences for a common ancestry of organisms, though indirect, is their fundamental similarity. In spite of many adaptive differences, protoplasm and cells and their manifestations of life—metabolism, growth, and reproduction—are essentially the same in all organisms.

### Lines of Indirect Evidence.

EVIDENCE FROM PALEONTOLOGY. Remains of previously existing organisms or any indications of their presence are *fossils*. These fossils exist in various types of rock and soil formations. Just as in a lake bottom, the mud on top has been most recently deposited, so in rock strata, the strata on top are more recent than those beneath if they have not been secondarily folded. With this in mind, the geologist is able to construct a chronological series of fossils, associating them with particular periods of geological time. Of course it has been known for a long time that fossils demonstrate the presence of many animals and plants in the past that no longer exist. When the fossils are arranged in a chronological series, faunal and floral changes are found that can only be explained logically by a series of progressive changes, viz., organic evolution. (Table 18.1.)

## Table 18.1. Geological Time Scale *

| ERA = time<br>GROUP = rocks | PERIOD = time<br>SYSTEMS = rocks | LIFE RECORD (FOSSILS)<br>BOTH ANIMALS AND PLANTS |
|---|---|---|
| CENOZOIC<br>Age of mammals<br>and modern flora | QUATERNARY | Periodic glaciation and origin of man (Pleistocene). The transformation of the apelike ancestor into man may have begun in the Pliocene. Culmination of mammals (Miocene). Rise of higher mammals (Oligocene). Vanishing of archaic mammals (Eocene). |
| | TERTIARY<br>upper<br>lower | |
| MESOZOIC<br>Age of reptiles | CRETACEOUS | Rise of the archaic mammals in the interval between the Mesozoic and the Tertiary. This Era is remarkable for the great development of the ammonites which became extinct at the end of the Cretaceous. The mollusks are more highly developed in this Era than in the preceding one. Culmination and extinction of most reptiles (Cretaceous). Rise of flowering plants (Comanchean); birds and flying reptiles (Jurassic); dinosaurs (Triassic). |
| | JURASSIC | |
| | TRIASSIC | |
| Upper<br>PALEOZOIC<br>Age of amphibians and lycopods | PERMIAN | Periodic glaciation and extinction of many Paleozoic groups during and after the Permian. Rise of modern insects, land vertebrates and ammonites (Permian); primitive reptiles and insects (Pennsylvanian); ancient sharks and echinoderms (Mississippian). |
| | CARBONIFEROUS | |
| Middle<br>PALEOZOIC<br>Age of fishes | DEVONIAN | First known land floras (Devonian) not very different from those of the Carboniferous. Earliest evidence of a terrestrial vertebrate in the form of a single footprint from the Devonian of Pennsylvania. Rise of lungfishes and scorpions (first terrestrial airbreathers) in the Silurian. |
| | SILURIAN | |
| Lower<br>PALEOZOIC<br>Age of higher<br>(shelly)<br>invertebrates | ORDOVICIAN | Rise of nautiloids, armored fishes, land plants and corals. Also the first evidence of colonial life (Ordovician). First known marine faunas; dominance of trilobites; rise of animals with hard shells or exoskeletons (Cambrian). |
| | CAMBRIAN | |
| PROTEROZOIC<br>Primordial life | PRECAMBRIAN | Fossils almost unknown except for a few problematical forms in the Proterozoic. Fossils unknown in the Archeozoic. |
| ARCHEOZOIC<br>Most ancient life | | |

NOTE: Geological time scales are constructed to show the oldest periods at the bottom and the youngest periods at the top. *To get the proper order and sequence of events, always read from the bottom to the top.*

* Reprinted by permission from *Geology* by Richard M. Field, Barnes & Noble, Inc.

EVIDENCES FROM COMPARATIVE MORPHOLOGY.

*Analogy and Homology.* Many rather unlike organisms have organs of similar function. If they are fundamentally unlike except in function, they are said to be *analogous*. Analogous structures indicate no close relationship except in habitat. (Examples: tail fin of lobster and flukes of whale, wings of fly and bird.) Organs fundamentally the same in structure, but perhaps modified for widely different functions, suggest a common plan that can be explained only through common ancestry. Thus, the arm of a man, the wing of a bird, the wing of a bat, the flipper of a seal, and the foreleg of a dog all have the same type of skeleton. Many of the bones correspond directly from one animal to another; all would, if it were not for the evident loss of certain ones. In plants, the scales of an ovulate pine cone correspond with the carpels of a lily, and the scales of the staminate cone are homologous with the stamens of the flower. This fundamental similarity is called *homology*. The criteria of homologies are, primarily, similarity in embryonic origin and, secondarily, similarity in structure. The presence of homologies, as brought out in studies of comparative morphology, can only be explained logically by a theory of organic evolution. (See Figs. 11.23 and 18.1.)

*Vestigial Structures.* Various organisms possess nonfunctional structures which, in other organisms, have essential functions. The caecum of the rabbit and other animals is homologous with the caecum and vermiform appendix of man. In man the structure is not only of no value but it does more harm than good, whereas in the rabbit it is a very important functional part of the digestive system. In man its presence may best be explained on the grounds that it is a structural vestige, something which functioned in man's ancestry but exists now only as a useless relic. The vestigial muscles at the base of the human ear, the caudal vertebrae (coccyx), and other human structures may be explained only in the same way. In plants, the scalelike leaves on the Indian pipe, a plant which has lost its chlorophyll and become saprophytic, are vestigial, as also are the stamens which in some flowers bear no anthers.

EVIDENCE FROM PHYSIOLOGY. Just as homologous structures exist, homologous functions also occur. It cannot be an accident, for example, that photosynthesis—with its remarkable function of carbohydrate synthesis—is so widely distributed through the plant world. Various types of chemical tests show close similarity between the body fluids of animals, serum precipitation tests on mammals indicating, for example, a much greater chemical kinship between man and apes than between man and

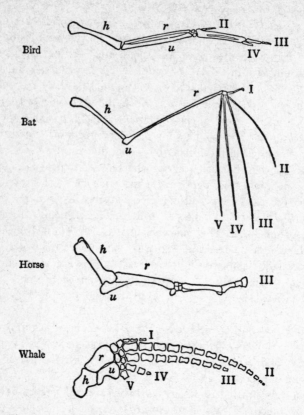

Fig. 18.1. Homologies of bones of the pectoral limbs in the bird and three mammals. Abbreviations: *h*, humerus; *r*, radius; *u*, ulna. Roman numerals refer to the elements of the primitive 5-digit appendage present in each. (Modified from various authors.)

swine. Another interesting line of chemical evidence is the remarkable similarity in proportions of constituent salts between vertebrates' blood and sea water. This supports the theory that the ancestors of all vertebrates were inhabitants of the sea.

EVIDENCE FROM EMBRYOLOGY. From the earliest stages of development remarkable similarities are found among organisms. In both plants and animals, for example, the formation of gametes is accompanied by a reduction in chromosome number. Higher plants, even such unlike ones as pine and lily, have embryos so much alike that the same terminology is used in describing their parts. Among animals, the stages of cleavage and blastula and gastrula formation are fundamentally the same in such dissimilar animals as starfish and frog, earthworm and man. The nearer the relationship of adult structures, too, the greater is the similarity in course of development.

This observation led to the formulation of a theory that, though doubtless largely true, has probably been too widely applied—the *biogenetic law*. It may be expressed: *Ontogeny repeats phylogeny;* or: The development of the individual recapitulates the development of the race. However, it may be more accurately stated: The development of an individual is similar to the embryonic development of its ancestral types.

Among vertebrates, all embryos pass through a stage in which gill-like structures and their associated blood vessels are present. In fishes, the condition is adult; gills function in some adult amphibia but only in the tadpole stage of higher forms; in reptiles, birds, and mammals these gill structures are never functional, but nevertheless are always present in the embryo. Evolution explains the presence of such an apparently unnecessary stage on the grounds that the embryonic gills of higher vertebrates are embryonic vestigial organs. This suggests that the ancestors of reptiles, birds, and mammals possessed gills. The heart shows similar relationship, the fish heart having one auricle and one ventricle, hearts of amphibians and reptiles having two auricles and one ventricle (partially divided, in reptiles), and hearts of birds and mammals having two auricles and two ventricles. In embryonic development, birds and mammals pass through *all* these stages, and in the order just given. Among plants illustrations are perhaps less apparent, but they occur: The fern spore germinates into a thalluslike (algalike) plant; the cycads, primitive gymnosperms, have fernlike motile sperm cells.

EVIDENCE FROM TAXONOMY. The principle of homology and the biogenetic law are the chief concepts used in classification. The evidences from taxonomy are, then, the evidences from comparative morphology and embryology.

EVIDENCE FROM GEOGRAPHICAL DISTRIBUTION. In groups of islands, the plants and animals of nearby islands are more alike than those of distant islands, whereas all organisms of such an island group show certain affinities with those of the nearest continental land mass. The assumption is that, with isolation due to the appearance of barriers, evolution has proceeded from a common starting type in gradually diverging lines. In conjunction with paleontology, good evidence for evolution may be deduced from the present and past distribution of camels and tapirs. The former are represented today by the true camels of the Old World and by the llama and its relatives in South America. Tapirs occur today only in South and Central America and in Malaya. Fossil camels and tapirs have, however, been found in the intervening territory. The natural assumption is that the forms occurring in the intervening territory have become extinct, leaving only the descendants in the extremes of the range. The existence of a land bridge between present Alaska and Siberia, and a milder climate there at that time than now, are requirements of the theory, but for both conditions there is abundant geological evidence. Based on evidence from fossils, maples probably originated in what is now northern Canada or Greenland; they are not now known in their ancestral home but are common in the more temperate northern sections of both Old and New Worlds—a discontinuous distribution. Many illustrations of this particular evidence for evolution are available from both plants and animals.

One bit of evidence for evolution comes from combining observations from distribution and taxonomy. The species of a major group in a particular region occur in different habitats and with different habits. (Several species of warblers can live in the same forest.) They are adapted to different environmental situations in such a way that the various related species do not compete with each other. It is as if a variety of favorable variations had been preserved in different, closely related genetic lines. Development of such a group of related forms is referred to as *adaptive radiation*. One of the best-known groups of forms related by adaptive radiation is the marsupials of Australia, which, in the absence of other types of mammals, have become squirrel-like, ratlike, marmotlike, antelopelike, and even wolflike.

## Lines of Direct Evidence.

EVIDENCE FROM GENETICS. Animal and plant breeders have long been able to develop domestic forms of desired characteristics by selective breeding; students of genetics, similarly, have been able to obtain evidence for evolution through controlled experiments and the study of mutations. Population genetics, the study of inheritance within interbreeding populations, supplies abundant evidence for changes of an evolutionary nature. At least one type of genetic experiment (a type of hybridization) has produced artificially a species that presumably arose in nature by the same method.

EVIDENCE FROM OBSERVATION IN NATURE. If large numbers of specimens of one or a few species are collected in a given locality, and the collection is repeated after a long interval of time, one may sometimes detect evolutionary changes which have taken place. Such changes were found by Crampton, studying land snails of the genus *Partula* from Moorea. His extremely large collections were separated by an interval of only fourteen years. Man's recent but extensive use of pesticides has been followed by the development of resistant strains of the pests involved. Modification of the environment has also influenced evolutionary change. For example, the increase in dark backgrounds caused by industrial fumes has resulted in a shift in English moth populations, darker forms becoming more frequent than lighter ones because of the selective advantage of protective coloration.

## EXAMPLES OF EVOLUTIONARY SERIES

**Evolution of the Horse.** The most ancient horse known (*Eohippus*) existed in the Eocene. It was about the size of a medium-sized dog, with somewhat the proportions of a modern horse. Its muzzle was shorter than that of a modern horse, however, and its teeth were low-crowned —better adapted for browsing than grazing. The forefeet bore four functional toes and the rudiment of the fifth; the hind feet had four toes—lacking the first digit. The ancestor of *Eohippus,* not yet found, probably had five functional toes on each foot. In subsequent evolution, the number of toes has been progressively reduced through the stage of three on each foot to the modern condition in which only the middle toe is present. (The second and fourth are, however, represented by the "splints.") In the skull, the muzzle has become progressively longer, and the teeth have developed high crowns with ridged surfaces— adapted for grazing. All along, the fossil series, which is very complete, shows gradual increase in size. Horses began their evolution in North America, but completed it in the Old World, becoming extinct on this continent and not returning until introduced by the first Spanish explorers.

**Evolution of Man.** Human fossils cannot be arranged in such a well-ordered series as can those of the horse. This is because early human remains have been rarer, and they must be distinguished from fossils of other types of contemporary primates and even from manlike types that early became extinct. Some fossil races have been known for a long time, particularly Neanderthal and Cro-Magnon man, but fossils of older races were not discovered in quantity until recent decades.

The oldest form probably in the line of human evolution is *Austra-lopithecus africanus,* discovered by Dart in South Africa. *Homo habilis,* discovered by Leakey in East Africa, is somewhat more advanced but is still considered by some to be "australopithecine." The next human fossils in chronology, including the Java man and the Pekin man, are now considered different subspecies of *Homo erectus.* The last of this type, Neanderthal man, disappeared about the time modern man, *Homo sapiens* (Cro-Magnon man), appeared. The earliest tool-making man probably lived two million years ago, in Africa, and man was present throughout many parts of the Old World a million years ago. (Other early hominoids, *Paranthropus* and *Zinjanthropus,* probably became extinct by the middle Pleistocene.) For a summary of significant fossil types of man see Table 18.2.

Human evolution, beginning in the early Pleistocene or earlier, has involved the following structural changes:

(1) Increase in brain size and development of a more rounded cranium (Fig. 18.2).

(2) Development of erect posture, with accompanying changes in pelvis and feet.

(3) Disappearance of prominent supraorbital ridges.

(4) Reduction of jaw size.

(5) Development of vertical face; formation of a chin.

(6) Reduction of canines, narrowing of incisors, flattening of cusps on molars.

## THEORIES OF THE METHOD OF ORGANIC EVOLUTION

Historically there have been three great evolutionary theories, but these are not mutually exclusive. The views of most modern biologists combine the second and third of these. There is no experimental evidence for the first theory—Lamarck's.

**The Theory of Lamarck—Inheritance of Acquired Characteristics.**

(1) Structural variations are acquired as a result of need. (There is no evidence in modern biology for this phase of the theory.)

(2) Use of a structure increases its size; failure to use it results in its atrophy or disappearance. This is the "principle of use and disuse." (In a general way, of course, there is some truth in this phase of the theory.)

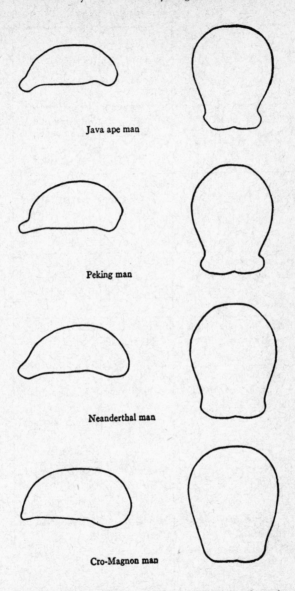

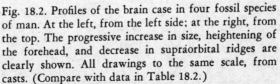

Fig. 18.2. Profiles of the brain case in four fossil species of man. At the left, from the left side; at the right, from the top. The progressive increase in size, heightening of the forehead, and decrease in supraorbital ridges are clearly shown. All drawings to the same scale, from casts. (Compare with data in Table 18.2.)

### Table 18.2. Fossil Types of Man

| Name and Locality | Geological Epoch | Nature of Fossils | Approximate Brain Volume (in cc) |
|---|---|---|---|
| *Australopithecus africanus* South Africa | Early Pleistocene | Skulls, skull fragments, and other skeletal parts | 550 |
| *Homo habilis* East Africa | Early Pleistocene | Remnants of seven skeletons, including parts of skulls | 680 |
| *Homo erectus erectus* ( = *Pithecanthropus erectus*) Java Man Java | Early Pleistocene | Skull fragments from four individuals | 870 |
| *Homo erectus heidelbergensis* Heidelberg Man Germany | Early Pleistocene | One mandible | ? |
| *Homo erectus pekinensis* (= *Sinanthropus pekinensis*) Pekin Man China | Middle Pleistocene | Fairly complete skeletal material from more than a dozen individuals | 1075 |
| *Homo soloensis* Ngandong Man Java | Middle Pleistocene | Skull fragments from eleven individuals | 1100 |
| *Homo erectus neanderthalensis* Neanderthal Man Europe, Asia, Africa | Middle and Late Pleistocene | Many skeletons | Up to 1800 (equal to modern man) |
| *Homo sapiens* Cro-Magnon Man France, Spain | Late Pleistocene | Several skeletons | Up to 1800 (equal to modern man) |

(3) These variations (now referred to as "acquired characteristics") are inherited. Thus the progeny have the advantage of the favorable adaptations acquired by their parents. (This phase of Lamarck's theory has no support from modern biology.)

**The Theory of Darwin—Natural Selection.**

(1) Organisms are prodigal in their production of offspring, far too many being produced to survive. These progeny are not alike; they show much fortuitous variation.

(2) This prodigal production results in competition or a struggle for existence—actually more of a struggle to escape being destroyed.

(3) Competition leads to natural selection of the most fit through death of those less fit to survive.

(4) The progeny of the organisms most fit to survive inherit the characteristics of their parents—namely those characteristics which have made their parents most fit.

Darwin did not understand the origin of inherited variations nor the mechanism of their transmission. His principle of natural selection is valid, however, as it operates on variations regardless of their method of origin. Following August Weismann we now know that evolutionary changes cannot occur unless they are incorporated in the germ plasm, which is the only continuity from one generation to the next (Fig. 18.3). Natural selection, operating on characters resulting from changes in the germ plasm (see next paragraph), is the basis for modern interpretations of the method of evolution.

**The Theory of De Vries—Mutations.** Evolution has not taken place through the accumulation of many small variations; it has been due to the appearance of a series of sudden changes in the germ plasm, *mutations*. These may be very pronounced or minor, but they are not equivalent to ordinary individual variations. De Vries's theory, as first presented, involved major changes only. Today we use De Vries's term, mutation, but apply it to inconspicuous as well as conspicuous variations that are inherited.

**Modifications and Corollaries of the Principal Theories.**

ISOLATION. As a factor in the development of new forms, geographic isolation has certainly played a great part. The formation of barriers within the range of a species prevents interbreeding between organisms of different localities; consequently, genetic peculiarities appearing in one region may be so magnified and concentrated that a form becomes distinctly different from its nearest relatives. If these genetic differences result in the development of new reproductive behavior, physiology, or morphology, they may effectively block further genetic exchange be-

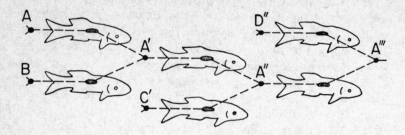

Fig. 18.3. Diagram illustrating the continuity of the germ plasm. There is no continuity of the somaplasm (body cells) between successive generations, but the reproductive cells are part of a continuous line connecting the most remote ancestors with all future decendants. All genetic and evolutionary changes must involve the germ plasm. In each generation a new source is added, but the continuity goes on, like a stream receiving many tributaries. (Reprinted by permission from *General Biology* by Gordon Alexander, published by Thomas Y. Crowell Company.)

tween populations that were once alike. Such blocks are called genetic or reproductive *isolating mechanisms*.

AGE AND AREA. According to the concept of Age and Area, organisms having widest geographical ranges are usually the oldest geologically. Hence, within a given genus the species having the widest ranges are probably most like the ancestral type, while those of very limited range are probably of recent origin.

HYBRIDIZATION. Crossbreeding between species may have produced much of the variation in nature—the heterozygotes introducing new combinations.

ORTHOGENESIS. Some students of evolution, particularly paleontologists, observe that fossils forming an evolutionary series indicate progress in a rather definite direction—the unsuccessful forms necessary to the theory of natural selection being absent. This theory that evolution proceeds in definite directions, from the directing tendency or limitations of internal structures, is *orthogenesis*.

**Present-day Views of the Method of Evolution.** Genetic changes occur spontaneously and relatively frequently. They may be gene mutations or chromosome mutations (see p. 251). These changes are distributed and preserved in a population by several factors. In a given population, if mating is at random, the ratios of all genes to each other tend to remain the same generation after generation (the *Hardy-*

*Weinberg Law*) unless (1) mutation rates of genes differ, (2) natural selection is involved (certain genes being selected for or against), (3) there is differential migration, or (4) genetic drift (variation due to chance in small populations) is involved. These four exceptions to the Hardy-Weinberg Law, plus the isolation of populations (implied in 3, above), are the important factors in evolution.

# *Chapter XIX*

# TAXONOMY

Taxonomy is the science of animal and plant classification. It has two aspects, the naming of each kind of organism (*nomenclature*) and the grouping of these to show relationships (*classification*). In order that species may be dealt with accurately each one must have one name and one name only. Classification of organisms is necessary for convenience, of course, its first purpose, but of equal importance now is the effort of the taxonomist to discover and indicate genetic relationships that exist between organisms of different species through evolution. Organisms classified in the same group are believed to be related closely by common descent.

## THE CONCEPT OF SPECIES

The working unit of the taxonomist is the species, usually thought of as having a real objective existence. No absolute criterion for recognizing a species as a unit has been found, however. Certain criteria may be suggested: A group of organisms sufficiently alike to have had the same parents belong to the same species; in other words, the extent of difference, morphological or physiological, is within the range of individual variability. Within a species of wide geographical range the variation may be so great that the extremes would not be considered of the same species did they not intergrade with each other. This is another test of species: Members of the same species interbreed freely, while those of different species ordinarily do not. This does not mean that hybridization between populations that are considered different species cannot occur. It can and does occur, but not frequently. It occurs between populations that have not yet evolved genetic isolating mechanisms. Our definition of a species must, therefore, be a dynamic one. It must allow for populations that are still in the process of evolving into distinct species. (If we had no problems in defining species we would have no evidence for organic evolution.)

## THE BINOMIAL SYSTEM OF NOMENCLATURE

Following the *Binomial System of Nomenclature,* the general adoption of which dates from Carolus Linnaeus (18th century), each organism is known by two names. These are: (1) the name of the genus to which it belongs, always written with the initial letter capitalized and (2) the specific name, its initial letter never capitalized. These two names constitute the scientific name. (Examples: scientific name of man, *Homo sapiens;* scientific name of dandelion, *Taraxicum officinale.*) It is customary to print scientific names in italics. The names are in Latin, for two reasons: (1) When the system was adopted, Latin was the international language of scientists and (2) since Latin is a "dead" language, its forms are not subject to change.

## CATEGORIES OF CLASSIFICATION

Although the species is the working unit of the biologist, for convenience in treating modifications within a species *subspecies* and *variety* names are sometimes employed. These constitute the third part of a trinomial. (Example: *Turdus migratorius propinquus* is the scientific name of the Western Robin; the eastern form is the one which corresponds to the originally described species; hence, the third or subspecific name is merely a repetition of the second—*Turdus migratorius migratorius.*) With this exception, the *species* is the smallest category in classification. Species form *genera;* a genus is combined with other genera to form a *family;* families form *orders;* orders form *classes;* and classes are combined into *phyla* (Fig. 19.1). Each succeeding category is more inclusive and larger than the preceding one. Thus a family may contain many genera, and itself be only one of several families in an order. In addition, large families may be divided into *subfamilies,* each containing several genera; or a family may be combined with other families in a *superfamily,* the order containing more than one of these. Other intermediate steps in classification may be inserted wherever they may clarify the complexities of animal and plant relationships. The criteria of these categories are not established; i.e., a beginner in

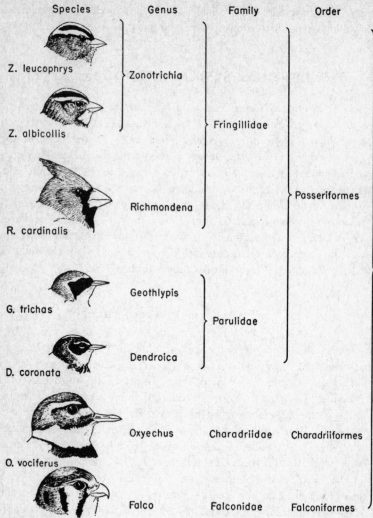

Fig. 19.1. The classification of each of the birds represented in the left column is indicated by columns of brackets to the right. Thus, the seven
species of birds are in six genera, four families, and three orders. All are
members of the Class Aves of the Phylum Chordata. The birds shown are,
from top to bottom, white-crowned sparrow, white-throated sparrow, cardinal, yellowthroat, myrtle warbler, killdeer, sparrow hawk. (Drawing
by Dominick D'Ostilio. Reprinted by permission from *General Biology* by
Gordon Alexander, published by Thomas Y. Crowell Company.)

taxonomy has no way of knowing just how different animals or plants must be from each other to constitute different genera rather than species, or families rather than genera. This puzzles even experts. However, the criteria in the animal or plant kingdom tend to be more or less uniform. These are based on the experiences and opinions of individual taxonomists—specialists in particular fields.

## EXAMPLES OF CLASSIFICATIONS OF PARTICULAR SPECIES

In the following examples, only major categories are given; there is no significance for the average biologist in th eintermediate steps omitted.

| | | |
|---|---|---|
| Common name ........ | American Elm | Red-tailed Hawk |
| Scientific name (Species) | *Ulmus americana* Linn. | *Buteo borealis* (Gmelin) |
| Kingdom ............. | Plant | Animal |
| Phylum .............. | Tracheophyta | Chordata |
| Subphylum ........... | Pteropsida | Vertebrata |
| Class ................ | Angiospermae | Aves |
| Order .............. | Urticales | Falconiformes |
| Family .............. | Ulmaceae | Accipitriidae |
| Genus ............... | *Ulmus* | *Buteo* |
| Specific name ........ | *americana* | *borealis* |

Note that the generic and specific names are singular in number while the names of all higher categories are plural. Thus one says "*Buteo borealis* is a hawk" but "The Falconiformes are the diurnal birds of prey."

## CERTAIN ESSENTIAL RULES OF NOMENCLATURE

**Publication.** A new species for which a name is being proposed must be described in connection with its name or the latter is not valid.

**Avoidance of Duplicate Names.** The scientific name of a new species must be such that the organism will not be confused with another. The name must differ from all other species names in the same genus; and there must be no duplicate generic names in the animal kingdom or in the plant kingdom.

**Priority.** The first name given a species is the accepted one if it is later described under another name; the later name for the same species is called a *synonym*. The name of the *author* (the person who gave the name) is ordinarily printed after the scientific name to avoid confusion over synonyms. (The name of Linnaeus is commonly abbreviated Linn., or simply L.; and other frequently occurring names are also abbreviated.) If subsequent study suggests that a species was placed in the wrong genus, or that the original genus should be subdivided, the genus name is corrected but the original species name is retained; and the original author's name is also retained, but in parentheses.

# Chapter XX

# ECOLOGY AND BIOGEOGRAPHY

There are two different approaches to the study of plant and animal distribution, the ecological and the geographical. In ecology, the biologist is concerned with the relations between organisms and their environment. The adaptations of organisms to their living and non-living environment constitute the heart of ecology; the ecologist who knows well these environmental relations will know where a particular kind of organism can live. In plant and animal geography, however, one studies the actual distribution in terms of existing land masses and the geological history of such land masses. The two approaches are overlapping, but the ecologist is interested primarily in where an organism can live and why, while the geographer is interested in where it actually does live and how it got there.

## ECOLOGY

**Subdivisions of Ecology.** Ecology may deal with the relations between one organism and its environment (*autecology*) or the relations of a group of organisms that occur together (*synecology*). Synecology includes the study of *populations, communities,* and *ecosystems.*

**Factors of the Environment.** The media in which organisms live (including specific nutrients), the substrata on which they live, the climate in all its aspects, and the presence of other organisms of the same and different species are all factors of the environment. We describe the media in terms of more or less stable physical and chemical properties and climate in terms of such changeable attributes as temperature and moisture.

NUTRIENTS. Various materials are taken from the environment by organisms and combined into organic compounds, using solar or metabolic energy, through the processes of photosynthesis and biosynthesis.

The elements used to produce organic compounds are required in a variety of forms by both plants and animals. Photosynthesis is the ultimate source of biological energy through the utilization of solar energy. Eventual breakdown of the organic compounds results in the cyclic return of nutrients to the environment for repeated use. This cyclic use and return of nutrients is considered an ecosystem characteristic (see below).

THE MEDIA AND SUBSTRATA. All living cells are found in an aqueous environment. As a result of protective coverings and multicellular structures, however, organisms live in air as well as in a wide variety of liquid environments. The major media-substrata types are water, air, soil, and the bodies of other organisms.

*Water.* Organisms have an aqueous composition and are assumed to have evolved in a watery environment (Chap. I). Water has a variety of ecological functions: as a nutrient, as an important environmental factor that controls the actions of other components of the environment, and as a medium in which a large variety of organisms live during part or all of their lives. Water acts as an insulator, preventing rapid temperature changes; it is also important as a solvent for many compounds, as a medium for ionization, and because of its high surface tension. Water's unusual property of maximum density at 4°C with expansion before freezing results in the formation of an ice insulation over lake surfaces; and in temperate regions, in fall and spring, increasing density of surface water of lakes results in "overturns" that aerate the deeper waters. In addition, of course, water has general influences on climatic patterns.

*Air.* Although some organisms can complete their life cycles suspended in water, air acts as a medium for organisms that spend part of their existence in contact with soil or water surfaces. The earth's atmosphere consists chiefly of nitrogen (about four-fifths), oxygen (about one-fifth), and variable amounts of water vapor, as well as small amounts of carbon dioxide and other compounds. The carbon dioxide is utilized as raw material in photosynthesis, and it is given off as a by-product of respiration. Similarly, the oxygen released in photosynthesis from the decomposition of water is used by organisms in aerobic respiration. Both these cyclic movements constitute parts of what is commonly called the carbon cycle. The nitrogen of the atmosphere is not available to most organisms directly, but some bacteria and blue-green algae are able to "fix" it, forming nitrates, in which form nitrogen is available to plants. A variety of organisms, particularly certain types of bacteria, use or make available for other organisms different types of nitrogen compounds; these function significantly in the nitrogen cycle, which involves the nitrogen of the atmosphere, nitrogen-containing compounds in living organisms, and intermediate products. The diffusion of various components of the air into the soil and water, and especially the reaction sequences produced by carbon dioxide in water (forming carbonic acid, as well as dissociation products), are important in controlling the availability of nutrients to organisms. Altitude also modifies the atmosphere as a medium; the density of air, and therefore the quantity (partial pressure) of its several constituents decreases with increasing altitude.

*Soil*. Although organisms found in the soil are technically either in air or in water spaces between soil particles, the soil composition has a pronounced influence on their occurrence and structure. The soil texture is important in relation to moisture conservation and availability, and it is also important for burrowing animals, such as earthworms, which may further modify soil composition and texture by their activity. Angle of slope and exposure to the sun are factors affecting drainage and the absorption of heat from the sun. (Soil factors are called *edaphic* factors.)

*Bodies of Other Organisms*. Internal parasites must be able to obtain oxygen under extremely unfavorable conditions, and, if they are parasites in the digestive tract, must be able to resist the action of digestive enzymes. They are protected from most enemies; and, in general, they have a more constant environment than do many free-living forms.

CLIMATE. Climate is sometimes defined as the "average of weather conditions." The elements that produce the varieties of weather and climate are primarily light, temperature, precipitation, humidity, and wind. These are modified by such features as latitude, altitude, the location and form of adjacent land and water masses, prevailing winds, and ocean currents. Although the different media in which organisms live are usually described as more or less stable they are actually subject to modification and control by the climate.

*Light*. The most important climatic factor is radiation from the sun. This includes the long rays of the infrared, visible light, and some ultraviolet radiation (short rays). Light primarily in the range of visible and ultraviolet light is the source of energy for various photochemical reactions (e.g., photosynthesis, which uses light energy in the visible spectrum), while heat energy is derived primarily from the long rays. Solar radiation heats the air, the soil, and water. It is responsible for evaporation of water. It is the cause of expansion of air, and thus, in conjunction with topography and the earth's rotation, is an important factor in major climatic patterns. Certain constituents of the atmosphere reduce transmission of ultraviolet light so it is significantly greater at high altitudes. The atmosphere is also responsible for what is called the "greenhouse effect": solar radiation absorbed by the earth's surface is reradiated as heat rays, which do not escape through the earth's atmosphere but conserve warmth near the earth's surface.

*Temperature.* The intensity aspect of heat energy is represented by temperature. Life as we know it is found within a specific temperature range—essentially the range of liquid water and aqueous solutions. Temperature, as a representation of the degree of kinetic energy, controls many basic reaction rates in organisms. Temperature sequences and cycles, produced by various climactic patterns, are also environmental stimuli that synchronize and control the activity of organisms.

*Other Factors.* Other climatic factors, especially water currents (e.g., the Gulf Stream), wind (local as well as regional patterns), and air moisture (cyclical and total precipitation, forms of precipitation, moisture deficit—in relation to evaporation rates) control biological activity within specific environments.

BIOTIC ENVIRONMENT. The presence of a particular species is the product of a variety of environmental factors, including food, protection, mates, and individuals of the same and different species. The living or *biotic environment* of an organism includes all other organisms, whether of the same species or of different species, that have any direct or indirect effect upon that organism.

**The Major Habitats.** Ecologists use the term *habitat* in two ways: (1) in reference to the specific place where a particular species naturally lives and (2) as a description of a region characterized by particular environmental factors and normally inhabited by specialized types of species. In this second sense, there are three major habitat types—the ocean, fresh water, and land—each with numerous subdivisions.

THE OCEAN. This is the habitat of *marine* organisms. (Note: the adjective "marine" applies to organisms of the sea, not to those of fresh water.) The study of the physical, chemical, and biological characteristics of the ocean is called *oceanography*.

*Characteristics.* The ocean is the most constant of all external environments. It is characterized by high salinity of little variability, the average range of salt concentration being from about 30 to 37 parts of salt per thousand. The temperature range is also relatively small—about 35°C.

There are regular ocean currents that affect the climate of the sea and adjacent land. The tidal fluctuations, twice daily, are of particular importance to organisms that live along the shore.

*Distribution of Marine Organisms.* Organisms living in the *intertidal zone* (between high and low tide) are adapted to alternate drying and wetting, to periodic feeding, to rapid temperature fluctuations (more extreme than in the ocean proper), and, in many places, to the impact of waves or a shifting substratum. Organisms living between the tide limits and just below low tide are said to be *littoral* in distribution. Organisms associated with the open ocean are called *pelagic,* those that float or drift with ocean currents being called *plankton* while active swimmers are referred to as *nekton.* Bottom or *benthic* organisms occur at all ocean depths. Those found below low-tide levels are subjected to fewer variables. The depths at which marine algae live depend primarily on the distance light penetrates, but organisms that depend upon food from higher levels occur at all depths in the ocean.

FRESHWATER HABITATS. The study of the physical, chemical, and biological characteristics of freshwater habitats is called *limnology.*

*Characteristics.* Because of the generally low salinity (therefore low osmotic pressure) of fresh water, organisms in this habitat must have a means of regulating the osmotic pressure within their bodies or must possess special organs for secreting excess water. Greater fluctuations in temperature and in concentrations of gases and ions in solution occur in fresh water than in the ocean. Other important differences from ocean habitats relate to the isolation of bodies of fresh water from each other, and to the presence of periodic factors such as rapid currents, high turbidity, stagnation, and drying-up.

*Distribution of Freshwater Organisms.* Distribution varies chiefly with the nature of freshwater bodies. In general we separate standing water (*lentic*) organisms from flowing water (*lotic*) organisms. Lentic organisms can be further subdivided into those of swamps and bogs and those of lakes, and lotic organisms into those of streams and rapid brooks.

LAND OR TERRESTRIAL HABITATS.

*Characteristics.* Terrestrial habitats experience the greatest fluctuations in climate, there being not only marked differences among regions but also variations in lengths of climatic periods. Factors such as temperature, moisture, and light vary greatly with seasons, with latitude, with altitude, and with topography. Soil and air temperatures vary enormously, the extremes being more than 120°C apart. Terrestrial habitats are often characterized by large plant forms that provide a matrix for other smaller organisms.

*Distribution of Terrestrial Organisms.* Subterranean organisms are least subject to variations in climate. Above-ground organisms may live in direct contact with the soil, as do higher plants, or they may live in various strata in the vegetation, e.g., on grasses, in shrubs, in trees. Though some animals may be aerial for long periods of time, all such are essentially terrestrial.

**Populations.** Some attributes of individuals are also attributes of a population, but a population has a set of structural and functional characteristics that are measurable only in population terms. Those characteristics that are a product of the collective members of the species living in a given area include: (1) *Distribution,* the area occupied by the population. (2) *Density,* the number of individuals per unit area. (3) *Pattern,* the spacing of individuals, whether aggregated, regular, random, or of an intermediate pattern in arrangement. (4) *Dispersal,* the movement of individuals into or out of the area occupied by the population. (5) *Age structure,* the composition of the population with reference to relative frequencies of different age classes. (6) *Natality,* the birth rate. (7) *Mortality,* the death rate. (8) *Growth form,* the trend in population size through time.

Populations have a potential for very rapid increase (*biotic potential*), which, because of environmental limitations (*environmental resistance*), is rarely attained. The generalized growth form of populations developing in a new area, whether under natural conditions or in the laboratory, tends to follow a characteristic pattern, which is called the sigmoid growth curve. In such a case the rapid increase is followed by a relatively stationary population size, the environmental *carrying capacity,* when the biotic potential is balanced by environmental resistance. This is an expression of Chapman's law, which states that population size is directly proportional to the biotic potential and inversely proportional to the environmental resistance. Natural populations, of course, fluctuate in size in a variety of cyclic and irregular ways, depending upon specific organism-environment relationships.

The relations among individuals within a population are *intraspecific.* These include: (1) Relations required for *reproduction.* (2) Relations of *assistance,* which include protection and rearing of young and cooperation in social organizations. (3) *Competition* for food, which may be passive as well as active. (4) *Hostility,* which may take the form of rivalry for territory (see Chap. XXI) or for mates, elimination of sick and injured, or the devouring of one mate by the other or of young by a parent.

**Interspecific Relationships.** Ecological relations between species are numerous and complex. They are the key to the existence or absence of different kinds of plants and animals in a single natural community. The term *symbiosis* is sometimes used to include all close relationships between species, whether harmful or beneficial. It was formerly used as equivalent to mutualism, but now it usually includes parasitism, commensalism, and mutualism (see next paragraph).

The various types of *interspecific relationships* include: (1) *Competition* for food or some required component of the environment. (With plants, this is to a considerable extent competition for space.) (2) *Predator-prey relations,* those between an organism that feeds on others and the organism fed upon. (3) *Parasite-host relations,* those between a parasite and the organism in which or on which it lives. The success of a parasite depends upon its ability to benefit from the host without destroying it. (4) *Commensalism,* relations that benefit and are required by one of the species without injury to the other. (5) *Mutualism,* relations of benefit to and required by both species. (6) *Protocooperation,* relations favorable to but not obligatory for both species. (7) *Slavery,* the relationship in which one species is captured by and used to serve another. The development of such relationships between species has produced a variable group of interacting systems. These are further complicated by the fact that any species pair may have a multiple group of relationships that vary with life history stages and environmental differences.

**The Community.** Within specific environments, naturally occurring groups of interacting populations form interrelated assemblages of organisms called *communities*. The constituent organisms of a community may be dependent upon similar climatic or edaphic conditions, but biotic interactions produce a complex assemblage called a *species network*. The various organisms in a community form *food chains,* which act as major energy flow channels from plants through animals. In any one community these food chains are interconnected in various ways, forming *food webs* (Fig. 20.1). The organisms in a community are spaced in relation to environmental factors and various population interactions. Such spacing may result in *stratification,* the establishment of layers of distinct kinds of organisms and types of organismal activity. For example, the organisms of forest soil, herb, shrub, and tree layers

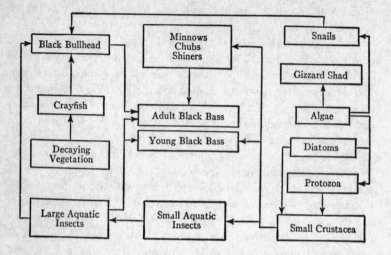

Fig. 20.1. Diagram illustrating a food web in a Kansas lake stocked with black bass. Arrows point from prey to predator. (Data from H. H. Hall.) Algae and diatoms are producers; the other organisms form various levels of consumers.

are of different species types. The largest or most numerous organisms in a community are called *dominants;* those having a major functional role are known as *key-industry organisms.* The transition zone between easily definable communities is called an *ecotone.* The condition of overlapping distributions, where gradual rather than pronounced shifts in community composition occur, is called a *continuum.*

**The Ecosystem.** It is obvious that the community and environment are not independent of each other. Together they form a complex system in which characteristic functions are performed by different species of organisms in different communities. The functional position of each species is called a *niche.* The whole system of organisms, in their dependence upon each other and on features of the nonliving environment, constitutes an *ecosystem.*

If the area is occupied by a self-sufficient community the ecosystem will have the following features: (1) Organisms that absorb light energy and utilize basic materials in the production of organic compounds. These (photosynthetic organisms) are the *producers.* (2) Organisms that obtain this organic material from the producers—for further biosynthesis and as a source of energy. These are *consumers.* The producers and consumers constitute a series of feeding levels called *trophic levels.* The consumers occupy two major trophic levels, the *primary consumers* or *herbivores* (animals feeding directly on producers) and the *secondary consumers* or *carnivores* (animals feeding on other animals). There may be several levels of carnivores, culminating in the top carnivore of the ecosystem; and, of course, some animals in their food habits are both herbivorous and carnivorous. A somewhat specialized trophic series is represented by the detritus feeders, the *decomposers,* which feed on discarded or dead organic material that may have been of producer or consumer origin. Decomposers are important as *transformers,* contributing to the renewed availability of various nutrients in the ecosystem. (3) The *nutrients* utilized in the construction of organic materials and in the maintenance of life are essential constituents of the ecosystem. These are continually being removed from resource pools during organic production and recycled to these pools through release during digestion and metabolism, and by activity of decomposers. Two of these *nutrient cycles,* the carbon cycle and nitrogen cycle, have been previously mentioned. Both habitat types and species forms influence modifications of the environment as well as rates of energy utilization and changes in nutrient cycles.

In comparing the productivity of various trophic levels one observes in some systems (e.g., the ocean) that the producers are the smallest and most numerous organisms and the final consumers are the largest and least numerous. This sequence forms a *pyramid of numbers,* the total number of organisms tending to be less in each succeeding trophic level. Even when the producers are larger and fewer than the herbivores—as in the case of trees fed upon by insects—the total mass decreases in successive trophic levels; this is a *pyramid of biomass.* The basic reason for these pyramids is the dissipation of solar energy as it is transferred through the trophic levels. This phenomenon of transfer of energy from sun to producers and through levels of consumers is called *energy flow.* The loss of some of the available energy in each succeeding step constitutes a *pyramid of energy* in the ecosystem. In terms of yield per unit area, man potentially obtains more when feeding on the lower trophic levels (plants) than when feeding on the higher trophic levels (animals). This is reflected in the differences in costs of food (e.g., bread vs. meat) and is indicated by the observation that plants form the staple foods in areas of high human density.

**Succession.** In each large area of the earth's surface, where environmental conditions tend to repeat themselves year after year, there develops a relatively stable and therefore generally characteristic group of plants and animals. This is referred to as a *climax community*. (Examples: the grasslands of the central states, with their prairie dogs and bison; the evergreen forest of central Canada, with its snowshoe rabbits and moose.) If such a climax community is destroyed in a given area, and that area is left undisturbed, the community will return to the climax condition through a directional development called *succession*.

Succession may proceed from a water environment to one of average moisture conditions, a *hydrosere* (Fig. 20.2), or from a dry environment to one of average moisture conditions, a *xerosere*. These are successions in time. One may also observe such successions in space, e.g., around the shores of a lake that is gradually filling up (Fig. 20.2), or along the edges of a disintegrating rock mass.

Although succession is community controlled, the physical environment establishes the rate of change and the final community form attained. As a result, the climatic differences in areas such as north- and south-facing slopes result in differences in succession and in the type of climax community. Along environmental gradients (e.g., altitudinal and latitudinal gradients), rate of succession and final community form show characteristic shifts. Because of this control by the environment, high-altitude communities are similar to those of high latitudes. The stability of the climax community is gained through increased control of the physical environment and is characterized by increased organic biomass (although the production per unit of biomass decreases), more complex food webs, and more complex and more nearly closed nutrient cycles. The communities maintained by man for biggest harvests are generally similar to successional communities of more rapid growth, and they often depend for stability on adjacent and less productive climax communities (e.g., irrigated farm lands depend upon forests on nearby mountains for water).

# BIOGEOGRAPHY

The ecologist considers an organism from the standpoint of its possible environments; a biogeographer considers that organism from the standpoint of where it occurs on the earth's surface and, historically, how it got there. Thus, an ecologist, had he known enough, could

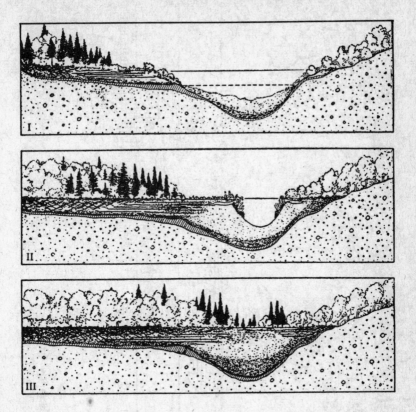

Fig. 20.2. Stages in a hydrosere succession: evolution of a peat bog. (After Dachnowski.) (Reprinted by permission from *Geology* by Richard M. Field, published by Barnes & Noble, Inc.)

have forecast the success of the European house sparrow in America; a biogeographer would have asked why it did not occur in America before it was introduced by man.

A species of plant or animal, under natural conditions, occurs over a part of the earth's surface called its *range*. This range may be *continuous*, or it may be interrupted by areas with a suitable habitat in which it does not occur; in the latter case the range is *discontinuous*. If a discontinuous range involves one or more small populations now separated from the main population with which they were once continuous these separated populations are said to be *relict* (an adjective). The interpretation of ranges is the province of the biogeographer.

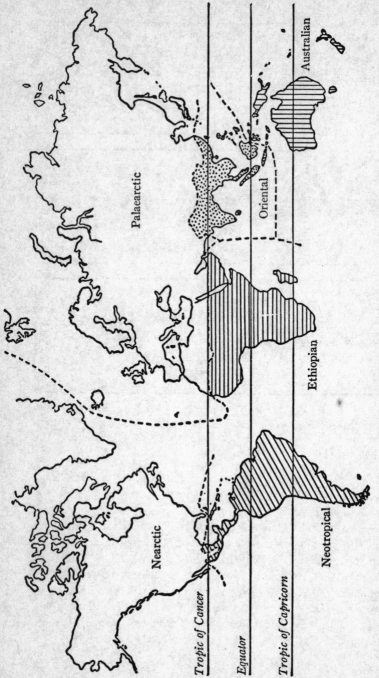

Fig. 20.3. Zoogeographical regions of the world according to A. R. Wallace. The Nearctic and Palaearctic are combined by many modern authors to constitute the Holarctic.

The biogeographer assumes that a species of plant or animal can occupy all the territory for which it is ecologically adapted if it can get there from its place of origin and if it has had time to do so. (Time is rarely an important factor because in the establishment of ranges we are dealing with geological time.) *Population pressure* is the major factor in causing a species to extend its range, and *barriers* are the major factors in limiting that range. The degree of movement possible to a species, its *vagility*, is important both in rate of movement and in the ability to overcome barriers. (A river, for example, may not be a barrier to a bird but may be to a mouse.) Some barriers block the most vagile of species. The changing of connections between large land masses (land bridges or isthmuses) or connections between large bodies of water (straits) are the most significant factors in range extension or restriction. Of major importance in explaining present distribution are the Bering land bridge between Asia and North America; various separations of South, Central, and North America; the long separation of the Australian region from elsewhere; connections between Africa and Asia other than those now existing across desert regions; possible former connections of South America and Africa, now separated by continental drift.

The most famous attempt to divide the world into biogeographic regions was that of Wallace, who based his divisions on the existence of families of vertebrates *endemic* to (confined to) different zoogeographic regions (Fig. 20.3). Other biogeographical classification plans may be partly ecological, as (1) the *life zones* of Merriam, based primarily on temperature zonation and secondarily on precipitation, or primarily ecological, as (2) *biomes,* which are regions typically characterized by dominant organisms. Such classification plans generally deal with smaller areas than geographical regions and the criteria used are necessarily predominantly ecological.

# Chapter XXI

# BEHAVIOR

The continued existence of an animal population depends upon the effective actions of the individual organisms. These actions, complex responses to environmental stimulation, directed toward keeping the animal alive and providing for reproduction, are aspects of *behavior*. Behavior is thus a component of irritability, which we have mentioned previously as a characteristic of living matter.

Behavior is dependent upon the ability of an organism to detect stimuli, transmit their effects, and produce an integrated response to them. Complex behavior patterns are limited to organisms with sense organs, nervous systems, and organs of locomotion, but primitive responses occur even in free-swimming unicellular organisms. The tropisms of plants are not considered behavior responses, though turgor movements (Chap. X) might be so considered. We shall limit our use of the term to responses of animals and free-swimming protista.

Because of the high mobility, pronounced size differences, and numerous adaptations of the sense organs, nervous system, and muscular system of animals, the study of animal behavior is a complex and distinctive branch of biology. (It is also referred to as *ethology*.) The study of animal behavior is of importance and interest to biologists because of its direct bearing on neurophysiology, endocrinology, and ecology. *Psychology,* dealing with the complexities of human behavior and the interrelations of man in society, is similarly based on the structure and function of the biological stimulus-response phenomena. The individual behavior of man merges into group behavior, as a result of man's complex social structure, and psychology bridges the gap between biology and the social sciences.

Behavioral responses are difficult to simplify because of the complex biological interactions that produce them. However, they may be divided into two main categories: (1) responses to the nonliving environment, such as the panting of a warm dog, winter hibernation of bats, and the migratory activity of a bird, and (2) responses to other organisms, which include the discrimination of different species and the recognition of differences among individuals of the same species. The evasive burst of movement and sudden halt of a rabbit that sees a predator, and the distinctive dive of a hawk that spots a rabbit are examples of responses to other species. Responses to other individuals of the same species may involve the recognition of sex, age, and physiological differences; examples of such responses include protection of a territory by a male bird when it sees another male, and the quite different courtship behavior when it spots a female.

In all cases, behavior patterns are adaptive, relating an organism to its nonliving and biotic environment by having survival value. The adaptive patterns are therefore preserved as part of the evolutionary heritage (or passed along by training), and they are as much a biological characteristic of an animal as its anatomy.

Individual behavior is an expression of the capabilities of the nervous system, through its control of the use of energy by the organism. Responses depend on the state of the sensory receptors and the effectors (muscles and glands). These in turn are controlled by the physiological state and previous environmental-behavioral sequences (e.g., training). Furthermore, the response to a particular stimulus may occur after a time lag. Because behavior patterns are related to increasing complexity, biological generalizations are often fallacious in this field. It is as erroneous to attempt to account for the behavior of complex animals on the basis of the behavior of the simplest ones as it is to describe the behavior of less complex types in terms of the human capacities of reasoning and emotion.

New behavioral properties characterize higher levels of complexity. The morphologically simpler organisms have relatively simple behavior potentials. In these, behavior is largely *innate—stereotyped,* genetically fixed, and often related to specific stimuli. In higher organisms behavioral responses become increasingly more intricate, drawn out in time, and potentially variable, involving to a greater and greater extent *learning,* which produces modifications in behavior on the basis of past experiences of the particular organism. The most complex behavior includes *reasoning,* responses involving the use of general concepts, but this is developed to only a limited extent in animals other than man. Thus, the pattern of behavior is simpler in flatworms than in arthropods; it is simpler in arthropods than in vertebrates; and, among vertebrates, it is simpler in fishes and amphibians than in birds and mammals.

In our discussion of behavior we will first consider the structural and functional bases. For convenience, we have divided the subsequent discussion of some representative aspects of behavior into that of individuals as opposed to that of groups, but the reader should realize that individual behavior and group behavior are closely interrelated and have, at times, overlapping components.

# STRUCTURAL AND FUNCTIONAL BASES OF BEHAVIOR

The simplest responses require: (1) Sensory structures that *detect* environmental changes, stimuli—e.g., light, pressure, chemical factors. (2) Structures that *conduct* impulses initiated by the stimuli—e.g., nerves. (3) Effectors, structures that *respond*—e.g., muscle cells. It is possible for all these to be incorporated in a single cell, as in ciliate protozoa, in which specialized organelles may function in these three different capacities; or as in coelenterates, in which cnidoblasts may be simultaneously receptors, conductors, and effectors; but the pattern is really exemplified by the reflex arc, involving three cells (Fig. 15.5) or more. With such a simple mechanism one certain stimulus elicits one particular response.

**Sensory Structures.** These structures range from apparently undifferentiated cell membranes to specialized organs with unusually low thresholds of sensitivity to particular types of stimuli. "Eyes" may be merely light-sensitive areas or complex structures such as the eyes of insects, cephalopod mollusks, and vertebrates that can interpret details of the environment in both form and color. Organs sensitive to mechanical impulses may be simple pressure organs, or they may detect sound waves in water or air, and even, as do the ears of insects and terrestrial vertebrates, discriminate between sounds of different pitch. Organs that detect chemical characteristics of the environment, particularly organs of smell, which detect volatile substances, are highly developed in insects and mammals. The three types of sense organs here mentioned— organs of sight, hearing, and smell—are the major sense organs involved in communication between animals. These and other receptors, such as those sensitive to changes in temperature or humidity, or those internal receptors that produce sensations such as hunger, also have specific effects on individual and group behavior.

**Conducting Systems.** In the simplest organisms, reception and conduction are combined in the same cell. In more advanced organisms, cells become specialized for different functions. Nerve cells, which are adapted for conduction, acquire increasingly complex interrelations as the nervous system evolves from simple to complex patterns. The nervous system may be: (1) A network, in primitive, radially symmetrical animals. (2) An axially oriented system, even in the most primitive bilaterally symmetrical animals. (3) A more advanced axial system in which nerve tracts are separated from concentrations of nerve cell bodies, or ganglia. (4) A system in which the degree of centralization (by ganglia) culminates in a specific, generally anterior area of control, a brain. (5) The most advanced systems, in which a spinal cord and brain are specialized for specific functions.

**Effectors.** Effectors consist of muscles and glands. The muscular response may involve movement of the whole organism or just a part of it. Flexible, jointed appendages, across which muscles operate, are associated with the more complex behavior patterns, and these, of course, vary with methods of locomotion—whether by swimming, running (walking), or flying. Muscular movements differ widely in their sequential relationships, duration, intensities, frequency, and over-all patterns of action. When a muscle contracts it stimulates nerves that may respond by terminating the initial action or by initiating further activity. Specific endocrine secretions are also involved in certain behavioral responses.

# PATTERNS OF INDIVIDUAL BEHAVIOR

Individual behavior can be classified as innate (stereotyped) behavior, learned behavior, and reasoning. The dividing line between what is innate and what is learned is not sharp, however, and most complex behavior patterns involve both.

**Innate Behavior.** Innate (stereotyped) behavior is considered primitive because it tends to be invariable, determined by genetically fixed, structural patterns that are not modifiable by experience and that are highly characteristic of individual species.

KINESES AND TAXES. Some of the most primitive innate responses are of whole organisms. A *kinesis* is a total body movement that differs in rate or activity (e.g., the number of turns) under variable conditions. Pill bugs in light tend to follow long movement paths, whereas in the dark they turn more often, following very short movement paths. A *taxis* involves body movements that result in the individual's orientation either toward (positive) or away from (negative) a stimulus. In a culture of *Euglena* illuminated on one side by moderate light, the organisms accumulate on the side near the source of light. This is because *Euglena* is positively phototaxic. Each individual turns toward the source of light because of the relation between the "eyespot" (stigma) and a sensitive receptor spot near the base of the flagellum.

The "avoiding reaction" of a paramecium is not a taxis in the accepted sense but is just as simple a response, and it does involve the whole organism. A paramecium swims continuously. When it encounters an obstruction it backs up a short distance before starting forward again. It then moves forward over a new path, at an angle of about 30° from the former path because of a combination of its backing up, its spiral shape, and the differential activity of its cilia. Subsequent repetition of the reaction eventually carries the paramecium into a path that avoids the obstacle. This behavior is sometimes called a "trial-and-error" response, but obviously it does not involve learning.

REFLEXES. Reflexes are inborn responses to particular stimuli, determined by inherent structure, and generally involving parts of, rather than whole, organisms. While essentially innate, reflexes vary from responses that are always involuntary—e.g., in man, constriction or enlargement of the pupil of the eye, and acceleration or depression of the rate of heart beat—to those that may occur either involuntarily or voluntarily—the knee jerk in man or scratch movements in the dog. They may involve very few sensory, conducting, and effector cells, as in the knee jerk, or a great many, as in the complicated process of breathing. Undoubtedly, however, reflex mechanisms provide the pattern on which more complex behavior is built.

INSTINCTIVE BEHAVIOR. This is unlearned, stereotyped behavior, characteristic of a species, and more complex than a single reflex act. It may be thought of as comprising a series of reflexes, the response to one generally initiating the next. Thus, instinctive behavior tends to be like a chain of reflexes. Insects, with extremely complex behavior patterns, illustrate instinctive behavior particularly well. A female solitary wasp, solitary except at mating and therefore incapable of being "taught" by its parents, goes through a complicated series of actions when it has reached the reproductive stage. It locates appropriate soil, digs a burrow, provisions it with caterpillars it has paralyzed by stinging, and finally lays an egg in the burrow. Each step is essentially a reflex act conditioned by the success of the preceding step. Such behavior is unlearned, and it will be repeated by females of the next generation in the same way and with no communication between generations. One aspect of this particular behavioral sequence is not instinctive, however. The wasp makes several trips to the burrow, instinctively closing it each time as it leaves, but it must find and recognize the place of the burrow each time it returns. This involves learning visual landmarks, as has been demonstrated experimentally. The whole procedure is complex, and it involves elements that are both innate and learned, the usual situation in complex behavior.

CYCLES, "BIOLOGICAL CLOCKS." Many behavioral responses are modified by daily or seasonal changes, and, while these are not limited to innate responses, they involve innate as well as modifiable behavior patterns. Rhythms of behavior correlated with daily changes, which are called *circadian* ("approximately daily") *rhythms,* may affect simple reflexes (such as regulation of the heart rate in man, or pigment changes of fiddler crabs) as well as the complicated behavior differences between diurnal and nocturnal animals. Circadian rhythms are obviously correlated with changes in light, temperature, and other variables between day and night, but they are sometimes so fixed in the constitution of an animal that changing the environmental conditions does not necessarily eliminate the rhythm. The organism may, under such conditions, behave as if it had an internal "clock." (That this is true to at least a limited extent even in man is evident in man's difficulty in adjusting immediately to a new local time schedule after a long east-west jet flight.)

Rhythms of other than daily periods occur, of course. Differences in behavior when the tide is in and when it is out are common among intertidal marine organisms. For example, some intertidal animals feed when the tide is in, others when the tide is out. Such organisms are said to exhibit *tidal rhythms.* A few animals are influenced by *lunar cycles,* certain marine worms (palolo worms) spawning only at the time of the full moon, though not at every full moon. Behavior cycles correlated with annual cycles (*seasonal cycles*) are particularly well developed. These are related to the temperature cycles of temperate regions and to moisture cycles of tropical monsoon climates. The major contrasts are between a time of general activity and a time of dormancy or inactivity. Reproductive activity tends to be during seasons of maximum warmth and maximum rainfall. Some organisms, birds in particular, relate to temperature cycles by migration, leaving the northern regions before winter. Bird migration, as well as other aspects of the annual cycle, appear to be controlled by complex internal mechanisms

—both neural and hormonal—that respond to external stimuli but, at the same time, reflect an internal rhythm, a sort of "internal calendar."

**Learned Behavior.** Behavior that does not follow innate, fixed patterns but that has been modified by experience is *learned*. Such behavior often leads to a goal, recognized on the basis of previous experience. The instinctive behavior of the wasp previously described appears to be motivated by a goal, too, maintenance of the species, but interpreting the behavior in this way is man's doing. It is anthropomorphic. This particular behavior pattern has undoubtedly been evolved and preserved through natural selection; its results are beneficial to the species, but the process is as mechanical as the activity of a computer. The term *motivation* is used, appropriately, for the significance of rewards and punishments used in establishing a learning process. The same term, or the word *drive,* is used also for what in some cases is a very simple process—the satisfaction of hunger or thirst (in this case innate behavior). But the ambiguity of the term is well illustrated when we talk of man's "drive" or "motivation" toward such goals as, for example, professional success or financial security. Here again, the pattern of behavior may be both innate and learned.

One special type of learning is the *conditioned reflex*. If a particular response, normally the result of a particular stimulus, can be induced by a different stimulus when this has repeatedly accompanied the normal stimulus, the response is said to be a conditioned reflex. (The reflex has become "conditioned" to the new stimulus.) Such a conditioned reflex was demonstrated by Pavlov when he showed that salivary secretion in dogs, ordinarily stimulated by meat, was induced solely by ringing a bell—after meat and the ringing of the bell had been used repeatedly as simultaneous stimuli. This type of conditioning is sometimes called *classical conditioning*. In an experiment like Pavlov's the stimuli involved no choice on the animal's part. One can, however, provide experiments in which the learning process is tested in animals by permitting choices between stimuli resulting in rewards and punishments. This type of conditioned response, called *instrumental conditioning,* as well as classical conditioning, suggests the close relationship between innate and learned behavior in certain situations. Conditioning illustrates the concept of *association,* some behaviorists preferring this term to conditioning.

Experiments have demonstrated that even flatworms can learn. In their learning, which is of the simplest type, they are able consistently to select the correct one of two alternative paths to food. This simplest type of learning appears first in the most primitive bilaterally symmetrical animals. In experiments to test learning ability, rewards or punishments or both supply the motivation. Learning is more rapid with both rewards for correct choices and punishment for incorrect ones. It is also more rapid in organisms of more complex nervous systems, but, though there is a direct correlation between greater learning and more complex systems, the actual mechanism of learning is still unknown.

**Reasoning.** The most complex behavior involves *insight,* comprehension of situations that are not duplicates of previous experiences but that can be analyzed by generalizing. Some higher animals show various degrees of insight, particularly in reaching goals by indirect means. This is only an elemental form of reasoning, however, and very little of what we study in animal behavior involves problems of reasoning. Only man thinks in abstract concepts, using symbols and language, and only man communicates readily, not only with other individuals by direct contact but, by writing, with future generations as well as with his contemporaries. Man's "behavioral" complexity is unique. Psychology deals with much of this; but so also, in the broadest sense, do anthropology and sociology—and even economics, history, philosophy, and other man-made and man-oriented fields of study.

# PATTERNS OF GROUP BEHAVIOR

No sexually reproducing animal is completely independent of others. If not parthenogenetic or a self-fertilizing hermaphrodite it must associate with at least one individual of the same species and of opposite sex at some time. This reproductive behavior may be complicated by courtship, home construction, and care of young. It may involve establishment of a defended territory and, therefore, may lead into certain types of conflict between individuals. In some animal societies it can be affected by dominance (see below), and in others it may be limited to individuals specialized for reproduction. Behavior related to self-preservation often involves group relations, too. The behavioral adaptations evolved by predators for finding and capturing their prey, and the responses evolved by prey organisms for escaping predators, are sometimes almost as complex as reproductive behavior. And an animal society often provides special behavioral benefits both in securing food and in providing protection from enemies.

### Reproductive Behavior.

MATING. The simplest group relations involved in reproduction are
the temporary associations of aquatic animals during spawning. These
may be as simple as "chance" associations of animals of opposite sex
that select the same habitat, response to the environment leading individuals of both sexes to occur together. Specific environmental stimuli,
as, for example, the full moon in the spawning of the palolo worm,
may result in reproductive behavior adapted to assuring a high percentage of fertile eggs. The swimming together of fishes when the milt
is shed over the spawn, or the clasping of the female frog or toad by the
male at the time the eggs are shed and are most easily fertilized, are
behavioral adaptations related to reproduction when fertilization is
external. In these cases of fishes and amphibians, identification of the
species and recognition of a member of the opposite sex by both
individuals are essential to reproductive success. Not only is visual
discrimination of size, form, and color involved; recognition of one
animal by another may include patterns of communication. These may
be display movements in fishes, and they may include production of
characteristic sound patterns in amphibians. All such behavior patterns
are species specific.

With internal fertilization the requirements of identification of one's
own species and recognition of the opposite sex are, of course, essential
for preservation of the race. Both identification and recognition are
accomplished by sight, by hearing, or by smell. Sight is an important
sense in almost all cases; it is particularly important among those animals
in which the sexes are strikingly different in form or color. Sexual
dimorphism is highly developed in arthropods and vertebrates. This
dimorphism may be accompanied, too, by distinct behavior patterns—
*prenuptial displays*—prior to actual mating. In spiders, the smaller male
may perform special movements of its appendages as it approaches the
female, these apparently communicating its identity as a sexual partner
rather than something to be attacked and eaten. The elaborate plumage
of certain types of male birds, e.g., upland game birds, is manipulated
in various ways during prenuptial display. In the sage grouse, visual
displays are supplemented by auditory ones—"booming" sounds. (But
bird songs have, in general, a different function. See Territorial Behavior, below.)

Internal factors, especially female hormone cycles, may be of major
importance in vertebrate reproductive behavior. Acceptance of the male
by the female mammal may be limited to a certain period of time, the

time of *estrus* or "heat." And in some rodents one copulation is accompanied by formation of a vaginal plug that precludes further copulation during that reproductive cycle. While in many animals the female may be inseminated by several males, a single mating is the pattern in many other species, even among arthropods.

TERRITORIAL BEHAVIOR. Selection of a suitable place in which mating occurs and in which the young develop may be little more than the animal's use of the habitat for which it is adapted. But many animals, including mollusks (octopus), some arthropods, and representatives from all groups of vertebrates, exhibit the series of behavior traits we collectively group under the concept *territorial behavior*. This involves (1) selection of the site, the *territory;* (2) advertisement of its location, generally by the male; and (3) defense of it against others of the same species, also generally by the male. In birds, where the phenomenon was first recognized, the territory is selected by the male bird at his arrival in the spring. He advertises its location by singing, which is the main function of bird song. This advertisement is mainly for the benefit of other males of the species. If the male acquires a mate he advertises and defends the territory through the nesting period. The characteristic feature of the aggression associated with territorial behavior is that it is expressed only toward others of the same species. Hence, several species of birds may have nests in the same tree, but there will be only one pair of robins or one pair of yellow warblers.

The aggressiveness of the male in defending his territory may include fighting but more often than not is simply chasing invaders—who are inclined to be submissive when not in their own territory. The male's aggressiveness is strongest in the center of his territory and decreases as he approaches its boundaries, which are determined by the places where his aggressiveness is matched by that of males in adjacent territories. All behavior involving conflict among animals, whether aggressive or submissive, is called *agonistic behavior*.

The defended territory is in many cases used both for breeding and for feeding. In other cases, as in colonial species, the defended breeding territory may be only a few square feet, while an extensive feeding area may be shared amicably by scores of individuals of the same species. The entire area over which an individual ranges is often called its *home range.* This may coincide with its defended territory, or it may include an extensive area in which conflicts between individuals do not occur.

There are several apparent advantages in territorial behavior, all of which may have contributed to the evolution and survival of such behavior. This spacing of individuals of a given species assures reduced competition for mates or food or both, it has a regulatory effect on population size by setting a limit on density, and there is some evidence that it reduces disease by reducing contacts within a population.

DOMINANCE. Some species of animals live in groups of various sizes, such as herds or flocks. These societies, if permanent, have other functions than reproduction, but some such associations (e.g., in the sage grouse) are temporary and are primarily reproductive. Whether temporary or permanent, however, they are usually characterized by the establishment of a hierarchy of *dominance* within the group. In chickens, this hierarchy is referred to as *peck order*. It is a sequence of dominance in which one cock is dominant, pecking and chasing all others, a second cock, submissive to the dominant cock, is dominant over the rest, and so on down, to the cock that is pecked by all. The hens in the flock also exhibit a hierarchy of dominance. One adaptive advantage of such dominance appears particularly evident in the sage grouse. In this species, the master cock on the strutting ground inseminates more hens than do any other cocks, thus assuring selection of the most vigorous male.

NEST BUILDING AND CARE OF YOUNG. Some of the most complicated behavior patterns are related to construction of nests and care of developing eggs (and, in insects, larvae and pupae). There may be care even when there is no "nest." The male toad fish "incubates" the eggs in his mouth. Other fishes prepare areas for the eggs in a stream bottom or in aquatic vegetation, and then they may physically protect these areas from intruders during the process of development. The male stickleback even aerates the eggs by fanning water over them. Many arthropods construct elaborate nests for eggs and young, e.g., the provisioned nests of wasps and bees. The behavior involved appears to be primarily instinctive. Just as instinctive are the nest-building activities of birds; these sometimes include inexplicable traits, such as the inclusion of a cast-off snake skin in the nests of the crested flycatcher. In other cases, an instinctive pattern may have been somewhat modified. The chimney swifts, now nesting chiefly in chimneys, originally nested in hollow trees; and certain species of woodpeckers find telephone poles acceptable substitutes for unharvested trees.

Feeding of young by parents is a highly developed behavior trait in birds and mammals. In this mutual behavior two separate groups of responses are distinguishable, *care-soliciting* and *care-giving*. In addition, there are species of birds in which an incubating bird is brought food by its mate, and there are species of mammals, particularly carnivores, in which the male brings food to the lactating female. Care of young, particularly in mammals, may continue through juvenile periods of considerable length; and, in herds of ungulates, the age groups constituting the herd may have quite different behavior patterns though occurring together.

**Nonreproductive Behavior.**

FEEDING. Group behavior, particularly in animal societies, may be well adapted to seeking out and finding food. The adaptive value of such behavior is well illustrated by what is called *imprinting*. Newly hatched chicks or ducklings follow the first large moving object they see; they are imprinted on it. What they follow is usually, of course, the mother bird that hatched the eggs, but this response—which occurs only if established shortly after hatching—can be experimentally established with human observers or even with mechanically moved objects. The adaptive value lies in the association of the young with an adult that leads them to food or water.

The advantage of group behavior is especially evident in animal associations that are primarily feeding groups. A flock of blackbirds that settles in a field probably disturbs more insects proportionately than single birds might, and observers have found clear evidence that more food is available to cattle egrets that associate themselves with water buffalo or cattle than to those that are solitary. The individual wolves in a pack cooperate in the hunting process, though *dominance*, another modifier of behavior, does affect the order in which these predators may feed on the kill. While dominance may be a factor in feeding behavior in an animal society, it is not necessarily the same thing as *leadership*. A flock of sheep may be led by the oldest ewe in the flock. It appears that leadership may have a more important role in food gathering, while the significance of dominance appears to be related chiefly to reproduction.

PROTECTION. Much of the most interesting behavior related to protection from enemies is in the highly developed series of responses of individual organisms. Of course they may fight or run away, or make use of a specialized threat, such as the odorous spray of the skunk or the fiercely aggressive pose of the harmless hognose snake. One of the most widespread reactions, however, is the *"freezing"* of an animal when a predator or potential predator appears. Being motionless is particularly effective when the individual has *protective coloration* or a pattern of camouflage. The motionless snow grouse or ptarmigan, white in winter and mottled like lichen-covered rocks in summer, is almost indistinguishable from its background by either bird or mammal predators. Many butterflies, moths, and other insects are colored in such a way that, though they may be conspicuous in flight, they blend into the background when at rest. Others, that are themselves edible, resemble in appearance distasteful or even harmful insects, e.g., a species of fly resembling a bumblebee. (This is *mimicry*.) The behavior of potential predators of these mimics is modified by learning, a predator that has sampled the noxious form learning to avoid its mimic as well.

Aggressive behavior against enemies is actually not common. Several bird species, however, of the group called kingbirds, are extremely effective in driving away even large predatory birds—crows and hawks —by attacks in which the predators are out-maneuvered by the speed of the smaller birds. In social groups of large mammals, e.g., bison and deer, large males may have protective roles. Their function is successful when the herd keeps together and, in danger, behind the large males.

MIGRATION. Some animals spend a good deal of their lives in places where they do not carry on reproduction. After years in the ocean, salmon migrate from the sea into freshwater streams to spawn, and the European eel, that has lived most of its life in freshwater streams of Europe, travels to the western Atlantic to spawn. These are one-way migrations, the adults dying after reproduction. Birds of the Northern Hemisphere migrate seasonally to warmer climates for the winter, returning north for the warm, breeding season. In both one- and two-way migrations, reproduction may be involved, but migration unrelated to reproduction also occurs (e.g., the southward migration of birds in the fall). In both types of migration, internal stimulating mechanisms are involved; and, in the case of birds, these are correlated with a seasonal cycle.

One aspect of the behavior of a migrating animal has provoked a wealth of study and conjecture in recent years. How do the animals find their way? They do follow similar routes year after year, birds nesting in the same area again and again after traveling many thousands of miles between nesting seasons. This *homing* ability is particularly highly developed in some types of sea birds that nest on oceanic islands but spend the rest of the time on the open ocean. Their navigation is of a much higher order than that of birds like "homing" pigeons, in which the ability is poorly developed in comparison with that of many wild birds. There is good evidence that birds determine directions by the arc of the sun in combination with recognition of time of day, and there is some evidence that they can even recognize directions from the position of the stars in a starry sky! They are also undoubtedly paramount observers of visual landmarks; therefore much of the short-distance migration, and that part of a long journey when they are nearing home, is explainable by remembrance of things previously seen.

# Part Five

# RELATIONS OF BIOLOGY TO MANKIND

# Chapter XXII

# ECONOMIC BIOLOGY

All human activities involving living organisms or their products include phases of economic biology. There is so much of biology in medicine, in agriculture and forestry, and in all phases of the conservation of a natural fauna and flora, that only a few suggestive details are here given.

## BIOLOGY IN MEDICINE

Man is an animal. The principles of structure that apply to lower animals apply equally to man. Through a study of the lower animals, and through the study of man as an animal, the underlying principles of human structure and function may best be understood. Most of the practice of medicine, therefore, either as it applies to man or to other animals (veterinary medicine), is based on biological principles. Similarly, the control of plant diseases, the study of which is called plant pathology, is chiefly by the application of biological principles.

**Disease.** Many organisms live at the expense of others; if they not only derive benefit from but actually do harm to their hosts, they are *pathogenic*. The abnormal conditions in an organism following or accompanying injury by a pathogenic organism is *disease*.

DISEASE-PRODUCING ORGANISMS. Diseases of plants are mostly caused by viruses, bacteria, and various fungi; but some insects and some parasitic worms produce symptoms of disease in plants. In animals, diseases are chiefly due to viruses, bacteria, protozoa, or parasitic worms. Viruses cause numerous diseases in both plants and animals. They cause influenza, measles, poliomyelitis, smallpox, and other diseases in man. Rickettsias cause typhus and Rocky Mountain spotted fever. Bacteria are the cause of diphtheria, tuberculosis, tetanus, typhoid fever, "strep throat," and many other pathological conditions in man. Protozoa are the cause of malaria (Fig. 6.4), sleeping sickness, and dysentery (amoebic). (Note: Sleeping sickness is African; American epidemics of the disease called "sleeping sickness" are of encephalitis, a virus disease.) Parasitic worms cause many human diseases, especially in the tropics —those affecting most individuals being hookworms. Hookworms belong to the phylum Nematoda. Many others of the same phylum and Platyhelminthes of the classes Trematoda (flukes) and Cestoda (tapeworms) cause human ills. (Fig. 11.5.)

TRANSMISSION OF DISEASE-PRODUCING ORGANISMS. Food or drink contaminated by fecal matter from those infected with intestinal parasites may carry organisms causing typhoid fever, dysentery, or Asiatic cholera. Insufficiently cooked food may contain young stages of flukes, tapeworms, or roundworms. Organisms may gain entrance through wounds—those causing tetanus, for example. Other organisms may enter the respiratory passages in the air. Some organisms are transferred by the bite of insects, ticks, or mites. These are referred to as *vectors* of the disease. In man, the organisms causing malaria, sleeping sickness, and typhus are carried by insects; that causing Rocky Mountain spotted fever is carried by a tick. The many viruses transmitted by arthropod vectors are now called *arboviruses,* "arthropod-borne viruses."

ANTIGEN-ANTIBODY REACTION. In general, organisms are sensitive to foreign proteins in the blood. When such are introduced into the human body, the blood develops some kind of resistance—varying with the nature of the protein. The foreign protein constitutes the *antigen;* the "neutralizing" agent in the blood is the *antibody.* If the foreign protein is soluble in the blood, it is *precipitated;* if it is insoluble (like a microorganism), different individuals are clumped together (*agglutinated*). Following this reaction, *phagocytosis* (the engulfing of the foreign material by white blood corpuscles) is greatly increased. Certain organisms, which may remain in only one part of the body, produce and discharge into the blood powerful poisons called *toxins* (e.g., diphtheria bacteria); an antibody, called under these circumstances an *antitoxin,* is developed, but in many cases too slowly to stay the course of the disease.

IMMUNITY. Some species of animals are immune to certain diseases; animals in general tend to resist later exposures to diseases from which they have recovered. These natural and acquired immunities are both believed due to the presence of appropriate antibodies in the blood, although there is such a thing as cellular immunity, as opposed to immunity due to specific antibodies. Immunity may be established in some cases by the injection of regulated amounts of antigen—e.g., by the injection of dead typhoid bacteria. In one well-known case, immunity against a mild disease, cowpox, confers immunity against a closely related but much more virulent disease, smallpox. Temporary immunity may be conferred by injection of an antibody.

(Note: In rare cases, in man, a violent disturbance may result from the second of a series of injections required in an immunization process. This is called *anaphylaxis*. There are methods by which it may be anticipated and avoided.)

CURE AND PREVENTION OF DISEASE. The trend of modern medicine is in the direction of prevention of disease through the establishment of immunity—*preventive medicine*. Diseases may rarely be cured except by some method based on antigen-antibody reactions. A few drugs are specifically beneficial, the best-known example being quinine in the cure of malaria. In recent decades, the sulfa drugs have become important. The most effective "drugs," however, are *antibiotics,* substances produced by fungi and bacteria and capable of preventing the growth of many pathogenic organisms. Penicillin and streptomycin are examples.

**Cancer.** The normal mature human body does not grow appreciably after it has reached maturity, and its rate of growth is not really very rapid after birth. This means that cell division is much less frequent in comparison with that in the rapidly growing embryo. Occasionally, however, cells in a particular part of the body may revert to an embryonic rate of division. A rapidly growing, spreading tissue is called *cancer,* and the growth is said to be *malignant*. A tissue growth that is growing at a slower rate and does not invade other tissues is said to be *benign,* and it may be called a *tumor*—though that term has a wider use. A cancer, by excessive growth, may injure the organ in which it occurs. If the malignant tissue is not removed by surgery or destroyed by radiation— the only present methods of "cure"—it may give off cells that are carried to other parts of the body and induce malignant growth elsewhere. This process is called *metastasis*. When metastases have been established, present-day medicine can do little. This is the reason for emphasis on early detection and treatment.

When the solution to the cause of cancer is found, it will undoubtedly be related in some way to the cellular metabolic mechanisms involved in mitosis. These mechanisms are apparently modified by certain

chemical compounds (collectively called *carcinogens*—"cancer producers") that may be present in our environment. There is evidence,
too, that some cancers are induced by or related to virus infection. And
there is good evidence that differences in susceptibility to cancer occur,
differences that are in part hereditary.

**Pollution in Relation to Man's Health.** Man's industrial advance,
coupled with increasing population density, has begun to produce
problems that affect his survival. Man is polluting his environment at a
greater rate than the pollutants can be broken down or carried away.
Pollution of lakes and rivers poses problems of water purification for
domestic use, but the most serious threats of pollution to human health
today are from air pollutants. (Pollution of soil and water by persistent
insecticides is considered later in this chapter.)

Waste products from combustion in automobile engines, manufactur-
ing plants, power plants, and even residential and commercial heating
plants are so modifying our environment in urban areas that measures
to reduce pollution must be taken for man to survive. Enormous
quantities of carbon monoxide and other injurious compounds are discharged from automobile exhausts, and some are modified into other
dangerous compounds in the air. Harmful air pollutants are also given
off by a variety of industrial plants. The most detrimental effects in
human health are on the respiratory system, an increase in emphysema
and other respiratory illnesses being correlated with increase in smog,
and other manifestations of air pollution. Some air pollutants, too, are
known to be carcinogenic. The effects are biological, but the cure for
air pollution depends upon new engineering and industrial practices,
chiefly (1) more complete combustion of fuel, and (2) uses of fuels
lacking or low in certain pollutants (e.g., sulfur). Both industry and
government are now working toward these ends.


# ORGANISMS OF VALUE TO MAN

**Plants.**

USES OF PLANTS. Plants serve man as sources of food, drugs, building
materials, paper, fuel (wood, coal, and probably natural oil), textiles,
rubber, ornament. Fundamentally, plants constitute our only direct
method of conserving the energy which we continuously receive from
the sun.

DOMESTICATION OF PLANTS. Three chief centers occurred in antiquity: Tropical America (Mexico to Peru), Eastern Mediterranean region (Egypt to the Tigris-Euphrates Valley), and the Orient (China). Many plants were cultivated by man several thousand years ago, e.g., in the Old World: tea, flax, grapes, figs, dates, apples, rice, wheat, sorghum; in the New World: sweet potato, tobacco, maize. Cotton was domesticated independently in the Old and New Worlds. Scientific selection of plants with a view to improving stock began over a century ago; other methods, e.g., hybridization, have developed principally since 1900.

**Animals.**

USES OF ANIMALS. These include food, sponges, hides and furs, textiles, beasts of burden, protection, hunting. Another function seldom considered is land-building: Large areas of land in the tropics rest on foundations of coral rock, made possible by the action of colonies of coral animals (and algae).

DOMESTICATION OF ANIMALS. Several kinds of animals were domesticated by the late Stone Age. Dogs were probably domesticated first, independently in several regions, and were the only domestic animal common to Old and New Worlds. In America, in addition, were domesticated the llama and alpaca, the guinea pig, and the turkey (the latter probably for ceremonial purposes). Indian horses of America were descendants of those brought over by the first Spaniards. Horses, asses, cattle of several types, pigs, goats, sheep, camels, elephants, reindeer, water buffalo, and cats were all domesticated in the Old World.

FISHERIES. The term "fisheries" is generally used to include all biological resources from water, both marine and fresh water. Fish furnish food, oil, fertilizer, and other commercial products. Some crustacea, e.g., lobsters, crabs, shrimps, are important foods. Many mollusks are sources of food, oysters and clams being of chief importance. Mother-of-pearl buttons formerly came from the shells of freshwater mussels. Pearls are formed within the shells of bivalve mollusks; these are now successfully cultured in Japan. (Culture pearls are real pearls, not imitations.) Whale oil, once much more important than now, may also be considered a fisheries product—although derived from a mammal. The propagation of freshwater fishes for game purposes is practiced on a large scale by state agencies. There are, of course, aquatic plant resources of commercial value too, especially in the sea, where algae are sources of food and food additives as well as gelatin and fertilizer.

# BIOLOGY IN AGRICULTURE, FORESTRY, FISHERIES

**Increasing Production.** Practical applications of biology in agriculture and related fields have involved basic knowledge particularly from physiology, genetics, and ecology. The role of nutrients in plant physiology has become the role of commercial fertilizers in agriculture—and even in fish culture, where fertilizers increase the food supply at lower trophic levels and thus result in more rapid fish growth. Genetics has made major contributions toward improvements in crop and meat production and in quality and quantity of milk and eggs. Examples of such contributions include the discovery that hybrid corn is a heavier producer than the homozygous strains from which it comes, that disease-resistant strains of wheat and other grains can be bred, that quality and quantity of orchard fruit can be improved by breeding, and that beef and milk in a herd of cattle can be improved by artificial insemination of cows with preserved semen from prize bulls. The contributions of ecology are particularly evident in fish culture (mentioned above), and in forestry, where an understanding of natural plant succession may govern the practices of harvesting trees for lumber and pulpwood.

**Control of Agricultural Pests.** The success of crop and orchard agriculture depends upon maintaining high densities of single species of plants. A field of wheat or corn is not a natural community, and the organisms that tend to convert it into a more natural community—insects that feed on the crops, or weeds that displace them—are undesirable from the point of view of the farmer. It is important that such pests be eliminated, if possible, or controlled in numbers if elimination is not possible. The methods of control are generally referred to as chemical and biological.

The speediest methods of control, and those that have been most used in agriculture, have been chemical—the use of insecticides and herbicides. These are necessary to control sudden and serious outbreaks, but the use of certain insecticides has now passed the point where the environment can absorb them. (Furthermore, genetic strains of harmful insects resistant to these insecticides have appeared.) DDT and other chlorinated hydrocarbons do not readily break down after application. They persist in the soil or in water for years, so they have become significant pollutants of rivers and estuaries. (Recent studies have found accumulations of up to 30 pounds of DDT per acre in estuary soil.) And, unfortunately, they are concentrated to much greater densities in tissues of animals in the environment. (Fishes concentrate DDT up to 20,000 times that in the surrounding water, and fish-eating birds have concentrations of up to 100,000 times that of the water in which their fish food lives.) The concentrations in animals may pass the lethal point, but, perhaps just as serious, are many sublethal effects on organisms at different levels in the food chain. The danger of the situation has been recognized in some quarters, and in some places the use of chlorinated hydrocarbons has been banned. Insecticides will continue to be necessary, but chemicals that do not persist will have to replace DDT and related types.

Biological control is generally thought of as the use of natural enemies, either predators or parasites, that may be introduced into the area in which the pest occurs. A more striking application of biological principles was the recent elimination of the screw-worm fly from Florida. This was done by releasing from airplanes, in numbers far exceeding the numbers of fertile males in the natural population, males that had been sterilized by radiation. These competed in mating with females of the natural population, so more females were inseminated by sterile males than by fertile ones. In each of several generations the population decreased markedly because of the increasing number of sterile eggs. This livestock pest was eliminated from Florida in a single season, and without any undesirable side effects such as damage to other organisms.

# BIOLOGY IN CONSERVATION

**The Principle of Natural Harvest.** Natural resources—minerals, soils, water, living organisms—are essential for man. Conservation means, primarily, their wise use. Some resources are nonrenewable (e.g., coal and oil), but living things are replaceable. Conservation of renewable resources involves primarily the principle of a *harvest,* annually or at appropriate intervals, no greater than production during the period. This principle governs the controlled cutting of trees in national forests, where trees are cut only at replacement rates. It governs the setting of federal regulations on hunting migratory ducks and geese. And it governs state legislation regulating the hunting of deer, elk, and other big game. In each case, conservation is the practice of permitting a harvest of plants or animals that can be replaced.

**Conservation of Natural Areas.** Wise use of natural resources also includes preservation of unspoiled natural areas. Increased urbanization has greatly increased, in recent years, man's demand for the possibility of seeing and visiting undisturbed wilderness—or something akin to it—but the principle of preserving natural areas was first recognized about a hundred years ago. In 1872, by act of Congress, the United States set aside Yellowstone National Park. This first national park has been followed by many others, some set aside even in recent years; there are now thirty-three, with others planned. In addition to these there are several national seashores and numerous national monuments. All such areas serve to preserve biological communities in essentially natural conditions, even though some are set aside primarily for other purposes, because exploitation of natural resources is prohibited.

*Chapter XXIII*

# SOCIAL IMPLICATIONS
# OF BIOLOGICAL THEORY

To the biologist, sociology may be defined as the ecology of man. The number of biological principles of significance for human society is great, and these are freely incorporated into the body of sociology by students of that field. Such problems as the following involve an application of biology in studies of human society: (1) the origin and evolution of the family and other social units; (2) growth and decline of populations; (3) the roles of heredity and environment in modifying human society; (4) the roles of geographic location and natural resources in determining cultural trends.

## BIOLOGICAL UNITS IN HUMAN SOCIETY

The individual is the unit in considering the psychological relations within society; the family is the unit in considering social relations from a genetic viewpoint. The integrative or uniting factor in the family is reproduction; the requirement that individuals of opposite sex associate together is, however, probably less important than the requirement that the young, because of their long period of infancy, be cared for by the parents. Although aggregations of lower animals which are primarily not for reproductive purposes do occur, these are more simple than the complex social organizations of man and the social insects. Complex animal societies are based on the reproductive function; other integrative factors, particularly in man, are of course involved, e.g., food, protection, force (as in slavery), education, religion, sentiment.

# THE EVOLUTION OF HUMAN SOCIETY— BIOLOGICAL ASPECTS

The same factors involved in the evolution of an individual species are involved in the evolution of a phase of human society. The process of socialization of the individual (produced by the factors of integration) may be considered the genetic unifying factor—the equivalent of heredity in organic evolution. Departures from the previous social type may develop (variations), which modify the nature of the integration. Selection of certain of these departures tends to encourage their development, just as selection encourages the development of a new plant or animal variation.

## THE ROLE OF COMPETITION

Competition as a *biological* factor is much less significant for modern than for primitive man. There is little of the "struggle for existence" that characterizes lower forms of life; it is rather a struggle for dominance. Competition is psychological, sociological, economic—not biological. Furthermore, *cooperation* is as important a factor in survival, and thus in natural selection, as is competition. However, cooperation is so marked in human society that those who are handicapped are frequently protected, and those who would be naturally eliminated as "unfit" in primitive societies are saved by modern medical science. Modern wars, in which death may come to any man, eliminate many of those who possess the most desirable traits, as well as large segments of the civilian population. If wars are "inevitable" then not only is the future of human society questionable, its very existence is at stake. Attempts to justify war on biological grounds (viz., that man as a territorial animal is innately aggressive) are irrelevant. If modern man is anything, he is unique among animals, not only in his ability to control his environment but also in his potential ability to apply intelligence to the elimination of undesirable behavior such as organized aggression, war.

## HUMAN POPULATION PROBLEMS

**World Population.** As Thomas Malthus pointed out early in the nineteenth century, the world population potentially increases in a geometric progression while the food-producing areas are limited. By equitable distribution we can still feed the world population, and we can augment the present food supply by more intensive agriculture in many countries and by use of incompletely exploited regions. But such increases in food supply will not long suffice if present rates of population increase continue. The available food will eventually set a limit on population density if man does not do so himself. The checks then will be those that Malthus pointed out—famine, disease, war. This gloomy prospect is the basis for the present awareness of human population problems.

**Limiting the Population.** The logical means of preventing the world population from becoming too large for its resources is the voluntary limitation of births—what is called in practically all parts of the world "family planning." Dependable and simple methods are available—contraceptive pills, widely used in the more affluent countries, and intrauterine devices, relatively inexpensive, and adaptable even for illiterates. Other methods, less dependable (e.g., the rhythm method) or more radical (e.g., surgical sterilization), are also widely practiced. But the desirability of family planning is not self-evident for many families and in many cultures. Widespread educational campaigns are necessary and are being conducted throughout the world. Some birthrates have begun to decline in recent years, perhaps indicating that the educational campaigns, along with the increasing availability of contraceptives, have begun to have an effect.

**Differential Birthrates.** In general, birthrates tend to be lower among those of superior education and better economic position than among individuals of little education and a low standard of living. This comparison holds for countries, too, those countries having lower average educational attainment and lower standards of living usually having higher birthrates. These differences would be accentuated if birth control were practiced in countries where education and living standards are at high levels and not elsewhere. The problems of population growth would then be compounded by a situation in which a disproportionate number of individuals need aid when population pressure is most acute. Population limitation is thus a world problem, calling for population regulation everywhere.

# IMPROVING MANKIND

**Heredity and Environment.** A generation ago biologists were inclined to consider heredity all important, while psychologists and sociologists believed that the environment is the essential element in shaping the individual. Today both are recognized as important. Hereditary traits are more significant in limiting the potential of an individual; the environment is more significant in determining his actual achievement. Improving mankind depends then on improving man's genetic composition and at the same time providing an increasingly better environment in which he can reach his potential.

**Eugenics.** The improvement of mankind by improving his heredity is *eugenics.* The goal may be achieved, theoretically, by selection. We know that all men are not born equal, that is, with the same physical and intellectual endowment. This difference at birth is primarily the result of differences in heredity. If future generations come from the best representatives of the present generation the average of human ability will be raised. The application of eugenics may be either negative, reducing the reproduction of the physically and intellectually unfit, or positive, encouraging the reproduction of individuals with the most desirable characteristics.

Negative eugenics takes several forms. One does not arbitrarily say that individuals with physical handicaps or of low intellectual attainment may have no children; the reason for these conditions may have been environmental, not hereditary. There are, however, human defects, both physical and mental, that are known to have an hereditary basis. To an increasing extent physicians are advising individuals with heritable defects to refrain from having children. And we do, to a certain extent, legally limit reproduction by individuals with inherited mental defects. (Institutional care of mental defectives has the effect of preventing their reproduction.) Compulsory sterilization of feeble-minded individuals has been authorized in many states. Compulsory sterilization is not extensively practiced, however. Even if it were, undesirable genes could not be eliminated in one generation unless they were dominant; and if they were recessive they would be reduced by only a small proportion of the population in a single generation.

The practice of positive eugenics has been, primarily, the encouragement of individuals of superior intellectual ability to have larger families. This, of course, produces a moral conflict if, at the same time, all parents are urged to limit family size. ("Who are the genetic elite?") The ideal situation would be for individuals of superior hereditary qualifications in all parts of the world to have large families, and for individuals with undesirable traits to have none; but to effect this distribution would not be easy. Some recent human geneticists have, however, gone far beyond this in their recommendations, suggesting the use of methods, already available, for selective human breeding. Human sperm, from selected donors of superior genetic qualifications, can be preserved and subsequently used in artificial insemination. Before long it will undoubtedly be possible to transplant human eggs, using one from a superior female in the uterus of a sterile mother or one with an heritable defect. While transplanting sperm and eggs into natural organs is not the same thing as producing "test-tube babies" it has the same implication—viz., producing babies "made to order"—and it can produce the same ethical dilemmas. The same problems will arise when man learns to substitute new genes for old ones; some biologists have suggested this possibility for the future. When man's capabilities reach this stage of the "brave new world," however, man may still decide that eugenics need not go so far as to convert human society into an agency primarily for producing prize human specimens.

**Euthenics.** The improvement of mankind by improving his environment is *euthenics*. One might think of this only as a temporary expedient, since improvement of an individual that is due to his environment is not biologically inherited. But many individuals of superior heredity are handicapped by an unfavorable environment, and an improvement in the latter releases hidden potentialities. More than that, a favorable environment helps man's general cultural advance. Human social evolution, because it involves communication between generations, actually does take place by "inheritance of acquired characteristics." What one generation learns can be passed on to the next, and man's cultural evolution is thus far more rapid than his biological evolution.

**Biology and Race Prejudice.** Human races differ from each other in various morphological and physiological features. Many of these differences are quite real. We recognize racial differences visually. On the other hand, we know of no racial differences in intellectual ability. Because of race prejudice, however, certain races in various parts of the world have been denied the same favorable environment—the same opportunities—reserved for the politically dominant race. Such denial produces and preserves an economic, social, and intellectual inferiority. This inferiority is clearly produced by the restrictive environment, for when individuals of all races are treated alike some in all races prove to be of superior ability and attainment. Just as there is no biological basis for race prejudice there is also no biological basis for objection to interracial marriage. The problems that result are social, not biological;

the progeny in racial hybridization are in no way inferior to either parent race.

## THE ROLE OF BIOLOGICAL DISCOVERIES IN THE DEVELOPMENT OF SOCIETY

Biological discoveries have contributed to social progress in the following ways: (1) The domestication of plants and animals by primitive man was an essential step in the establishment of more or less sedentary groups. (2) The discoveries of medicine have prolonged life and removed discomforts which formerly handicapped man. (3) Scientific agriculture has, by improving the breeds of domestic plants and animals and by combating their enemies, increased the food supply and therefore the wealth of the world. (4) Biology now proposes that the discoveries in the domain of human biology be applied to the improvement of mankind.

*Chapter XXIV*

# THE HISTORY OF BIOLOGY

The beginnings of biology are lost in antiquity. Man began as a hunter, undoubtedly, and his occupation forced him into a knowledge of nature. The extent of his knowledge is reflected in the cave paintings and sculptures left by the Cro-Magnon race. Domestication of plants and animals was followed by a still more intimate acquaintance with nature. Crude medical practice involved a certain amount of human anatomical knowledge, just as use of animals for food gave primitive man a certain superficial familiarity with their structures. Scientific biology began with Greek civilization.

## CHRONOLOGICAL SUMMARY, BY PERIODS

### Biology in the Ancient Historic World.

BIOLOGY IN ANCIENT BABYLON, EGYPT, ISRAEL, AND THE FAR EAST. In Babylon, the priesthood knew something of anatomy (particularly of sacrificial animals) and medicine, the latter unfortunately involved in astrology, however. The medicine of ancient Egypt was more practical, being based chiefly on knowledge of the human body. The ancient Jews contributed to our modern conceptions of hygiene through their laws. In the Far East, the knowledge of nature and medicine never got far beyond the primitive stage.

BIOLOGY IN ANCIENT GREECE. Scientific medicine was founded by Hippocrates (*c.* 460–377 B.C.); his method was empirical. The ethics associated with the profession date from him or earlier. Aristotle (384–322 B.C.) was the first great organizer of biological knowledge, the "Father of Biology." Theophrastus (*c.* 371–287 B.C.), his successor, was the "Father of Botany." Greek tradition continued in the Museum of Alexandria until about 30 B.C., but under Roman rule steadily declined.

BIOLOGY UNDER THE ROMAN EMPIRE. A work for plant identification appeared in the first century A.D. (by Dioscorides), and the first botanical drawings appeared. Some progress was made in anatomy and physiology, particularly by Galen (A.D. 130–200), the greatest physiologist of antiquity.

### Biology in the Middle Ages.

THE DARK AGES IN EUROPE. From about the time of the death of Galen until the thirteenth century, works of Greeks and Romans were recopied, both text and drawings, with no recourse to sources in nature. They progressively acquired more and more errors.

THE ARABIAN EMPIRE. During the Dark Ages in Europe, the Arabs not only preserved the finest of ancient Greek science but added some contributions of their own.

THE THIRTEENTH-CENTURY REVIVAL. Beginning somewhat earlier, but reaching a climax in the thirteenth century, Europeans began to translate into Latin the scientific works available to them in Arabic. In this way Aristotle's works reappeared.

### Biology during the Renaissance.
The thirteenth-century revival expanded into the Renaissance. Art became more naturalistic and involved a thorough study of anatomy (e.g., by Leonardo da Vinci, 1452–1519). In 1530 appeared the first of a series of *printed* and illustrated works on plants (herbals). In 1543 Vesalius's great work on human anatomy was published, and natural histories of animals appeared in the same century. In 1628 Harvey's work on the circulation of the blood was printed.

### Biology in the Seventeenth and Eighteenth Centuries.
The attitude of scientific men began to change—they looked to nature itself for information. The period of geographic exploration expanded their vision; the invention of the microscope intensified it. The first scientific societies and journals were founded. Detailed studies on the anatomy of small organisms and parts of larger ones were pursued, and cells were discovered and named (Hooke, Malpighi, Grew, Swammerdam, Leeuwenhoek). Schemes of plant and animal classification were devised (Ray, Linnaeus) following the stimulus of explorations.

### Nineteenth-century Biology.

MORPHOLOGY. Anatomy became not only more detailed, but comparative. Cuvier (1769–1832) founded comparative anatomy; he studied extinct as well as modern vertebrates. Lamarck (1744–1829), his contemporary, best known today for his theory of evolution, was a thorough student of plants and animals. Bichat (1771–1802) made a classification of human tissues. The cell theory was propounded in about 1810 (Chap. III). The doctrine that protoplasm is a universal characteristic of life was accepted within another thirty years. Hofmeister (1824–1877) demonstrated the fundamental similarities between higher and lower plants. Studies on mitosis, the history of germ cells, fertilization, and embryology characterized the last of the century.

PHYSIOLOGY. At the beginning of the century, the discoveries of the identity of combustion and respiration and of the formation of oxygen by green plants (Lavoisier and Priestley) gave an impulse to the application of chemical knowledge to organisms. The first organic compound to be synthesized, urea, was prepared by Wöhler and was followed rapidly by many others. Liebig (1803–1873) applied chemistry to living phenomena, especially in plants. Claude Bernard (1813–1878) was the great chemical physiologist of the century. The germ theory of disease was developed, and biological control of disease begun (Pasteur, Koch, Lister). Johannes Müller (1801–1858), extremely versatile, applied comparative anatomy, chemistry, and physics in physiology. Ludwig (1816–1895) also approached physiology from the standpoint of physics; he invented some of the most widely used laboratory apparatus of today.

EVOLUTION. Lamarck's doctrine of use and disuse, with inheritance of acquired characteristics, came at the beginning of the century. It was opposed by Cuvier, whose influence was great, with the result that it found little favor. Evolution was not widely accepted until after 1859, in which year was published *The Origin of Species,* by Charles Darwin (1809–1882). August Weismann (1834–1914) brought out, toward the close of the century, the importance of the germ plasm and the noninheritance of acquired characteristics.

GENETICS. The experiments of Gregor Mendel (1822–1884), performed in the 1860's, were not appreciated until 1900. Statistical methods were applied by Francis Galton (1822–1911) to studies of human inheritance.

**Twentieth-century Biology.** The present century has been characterized by a more intensive application of experimental methods. The application of physical chemistry in biology has become prominent; other aspects have attained considerable recognition—e.g., ecology. But the twentieth century has been the period of greatest advance in genetics and cell physiology. The mutation theory was proposed in 1901 by Hugo de Vries and marked achievements have come in the study of the mechanism of heredity, particularly by T. H. Morgan and his associates. In recent decades, biochemists have clarified many details of cellular metabolism, and electron microscopists have described the minute structure of cells. Their discoveries have involved the enzyme systems of respiration and photosynthesis, the compounds used in energy mobilization, the chemical nature of proteins, the roles of nucleic acids in cells, and the genetic code, together constituting what has recently been called *molecular biology*.

# BRIEF BIOGRAPHIES OF LEADING BIOLOGISTS

**Aristotle** (384–322 B.C.). The "Father of Biology." He was tutor of Alexander the Great and founder of the Lyceum at Athens. Aristotle organized the knowledge of his period. His chief contributions were on the natural history, anatomy, and reproduction of animals and on the nature of life.

**Theophrastus** (*c.* 371–287 B.C.). The "Father of Botany." Theophrastus was the successor of Aristotle in the Lyceum—not a thinker of equal rank, however. He left us the most extensive treatises on plants of the ancient world.

**Andreas Vesalius** (A.D. 1514–1564). Vesalius was a Belgian, who became professor of anatomy at Padua, Italy. In 1543 he published the first scientific treatise on human anatomy, beautifully illustrated.

**William Harvey** (1578–1657). English physician, educated at Padua, Italy. Harvey published the first accurate account of the course of the blood in the human body (1628); his embryological studies emphasized the origin of both viviparous and oviparous forms from the egg.

**Carolus Linnaeus—Carl von Linné** (1707–1778). Swedish botanist and zoologist. He established on a firm basis the binomial system of nomenclature—in the 1750's.

**Antoine Lavoisier** (1743–1794). French chemist. Established the fact that carbon dioxide and water are the end products of respiration as well as of ordinary combustion.

**Justus von Liebig** (1803–1873). German chemist. His recognition of the origin and role of organic compounds in living organisms established recognition of the carbon and nitrogen cycles.

**Charles Darwin** (1809–1882). He was naturalist of the *Beagle* on a trip around the world, 1831–36. On his return he established a notebook on the change of species. *The Origin of Species* was published in 1859—in its effect on human thought the most significant book of the century. Darwin published many other works, not all of which were in the field of evolution.

**Claude Bernard** (1813–1878). French physiologist. His discoveries in physiology included many significant details (e.g., glycogen storage, digestive functions of the pancreas, nervous control of blood pressure) but he is remembered chiefly for his emphasis on critical and objective methods of physiological research.

**Wilhelm Hofmeister** (1824–1877). German botanist, self-educated. He was the first to describe fertilization and embryo formation in higher plants. He demonstrated the wide occurrence of the alternation of generations in plants; and he pointed out the fundamental relationships of all plants.

**Gregor Mendel** (1822–1884). Monk, and later abbot, of a monastery in Austria. His most important contribution was the discovery of the laws of inheritance now called Mendel's laws.

**Louis Pasteur** (1822–1895). French chemist and bacteriologist. Simultaneously with Robert Koch, a German, Pasteur demonstrated the bacterial origin of disease. He developed a method of attenuating (weakening) a disease-producing organism so that it could be used in developing immunity. He established inoculation against rabies or hydrophobia. In his memory there now exist many Pasteur Institutes throughout the world.

**August Weismann** (1834–1914). German zoologist. Weismann proposed the theory of the continuity of the germ plasm. He made the first important scientific denial of the inheritance of acquired characteristics.

**Hugo de Vries** (1848–1935). Dutch botanist. His early experiments formed a basis for the theory of electrolytic dissociation. He was founder of the *mutation theory* of evolution.

**Theobald Smith** (1859–1934). American parasitologist and immunologist. Among many other discoveries in medicine, he was first to demonstrate experimentally the transmission of a disease-producing organism by an arthropod (Texas cattle fever transmitted by a tick, demonstrated in 1889).

**Thomas Hunt Morgan** (1866–1945). American zoologist, Nobel Prize winner. His experiments with *Drosophila,* coupled with those of his students and associates, were the basis for the gene theory of heredity.

**Hans Spemann** (1869–1941). German embryologist, Nobel Prize winner, whose application of experimental techniques in animal embryology resulted in the theory of embryonic induction. He introduced the term "organizer" in 1921.

**Clarence Erwin McClung** (1870–1946). American zoologist, who discovered the chromosome mechanism of sex determination (in grasshoppers) in 1901.

(Note: The following twentieth-century biologists have all made significant contributions. The list is representative rather than comprehensive, however, and is weighted in its selection of contributors in modern molecular biology. The biologists are listed in order of year of birth, not in the order of their major achievements.)

**Charles S. Elton** (b. 1900). British biologist. Emphasized food chain as factor in community integration; made other contributions in community and population ecology.

**Hans Krebs** (b. 1900). German-born, British biochemist, Nobel Prize winner for his discovery of important steps in the process of cell respiration. The citric acid cycle is named for him.

**George W. Beadle** (b. 1903). American biologist, Nobel Prize winner, who, with Edward L. Tatum (b. 1909), was responsible for the one gene-one enzyme theory in genetics.

**Konrad Lorenz** (b. 1903). Austrian-born, German biologist, who, with Nikolaas Tinbergen (b. 1907), has made important contributions to the organization of a science of ethology (animal behavior).

**Wendell Stanley** (b. 1904). American biochemist, Nobel Prize winner, who first crystallized a virus, and who has subsequently made many contributions to our knowledge of virus structure and function.

**Arthur Kornberg** (b. 1918). American biochemist, who was awarded the Nobel Prize in 1959 for his discovery of DNA polymerase, has recently (1967), with his associates, accomplished the synthesis of active DNA.

**Marshall W. Nirenberg** (b. 1927). American biochemist, who, with Heinrich Matthaei, determined in 1961 the basis for the genetic code by demonstrating the relation between a codon (UUU) and a particular amino acid (phenylalanine).

**James D. Watson** (b. 1928). American biologist, Nobel Prize winner, who, working in England with Francis H. C. Crick (b. 1916), proposed (in 1953) the accepted structure of the DNA molecule.

# Chapter XXV

# PHILOSOPHY AND BIOLOGY

Ostensibly or otherwise, a biologist believes in the independent existence of a world of objects which he, through his sense organs, perceives. Objects in that world, which exhibit a certain group of properties (life), are his province. Given the same instruments, all normal human beings are able to find the same properties in an external object. This uniformity of observations by different individuals means to the biologist that the objects exist apart from the minds that perceive them. Consequently, a biologist does not believe that mind is the only reality. Modern scientists have been handicapped by the view of Descartes that knowledge of one's self is prior to other knowledge. As a matter of fact, we do not know in full the relations between body and mind. The validity of knowledge, however, is not dependent on our understanding of how we know.

## LIFE

No simple criterion makes it possible to distinguish between living and nonliving matter. The complexity of life and its essentially mysterious nature make it a subject for conjecture.

**Mechanism and Vitalism.** In general, there are two antithetic approaches to life. The *mechanist* considers a living organism a machine, whose parts and their interactions obey the same physical and chemical laws which we know in the nonliving world. The *vitalist* believes that the ordinary physical and chemical laws are insufficient to explain life—that the condition of being alive is due to some other factor about which we know nothing, but whose existence we must assume. The former viewpoint has, of course, been the fruitful one in scientific research. That does not prove that life is a mechanism, however, but it does suggest that the practical or methodological approach involves an assumption that the known laws of chemistry and physics do explain living matter. Hence, modern biologists are mechanists in experimental approach, regardless of their fundamental convictions.

**The Organismic Hypothesis.** Certain biologists, and more particularly psychologists, have suggested that the activity or behavior of an organism is not the sum of the actions or functions of its separate parts, but that, as a dynamic whole, it transcends in behavior these separate parts. In other words, the whole is not functionally the sum of its parts. Such a view is the *organismic hypothesis.* According to the opposite hypothesis, the *atomistic,* the behavior of the organism is merely the sum of its separate functions. Followers of either hypothesis may be mechanistic in their general interpretation of life.

**Origin of Life.** Since the biologist has been unable to demonstrate the spontaneous origin of living matter, he has been inclined to ignore theories of the origin of life as outside his field. His working hypothesis of mechanism, however, implies that living matter has in some way been derived from nonliving. And there is now experimental evidence that complex organic compounds are formed spontaneously from the inorganic compounds that were present in the primitive earth's atmosphere. From these organic compounds it is possible to postulate formation of the first compounds that behaved with the properties of living matter. Henderson, some years ago, emphasized the reciprocal relationship between environment and organism, believing that the fitness of the cosmos for life suggests that both inorganic and organic evolution are essentially one, and that the universe is *biocentric,* viz., that its central and most fundamental fact is life, not nonliving matter.

**Evolution.** Various biologists and philosophers have suggested the presence of a directing or controlling force, external or internal, in evolution. Modern theories of evolution find this view irrelevant and unnecessary, however; it is now only of historical interest.


# SCIENCE AND RELIGION

Many unfortunate controversies between defenders of organized religion and scientists have arisen during the course of history. They have been, however, in final analysis, based on superficials—either of faith or of knowledge. The two provinces represent different viewpoints, but not necessarily opposite ones. Most controversy has been over the matter of organic evolution, particularly as applied to man.

Opponents of the evolutionary doctrine, none of whom is a trained biologist, have in common one viewpoint which is not consistent with the discoveries of biology. That viewpoint is a belief in the literal inspiration of the Bible and in the strict truth of every word in the particular translation they follow. Of course such a view is inconsistent with the findings of biology, but it is equally inconsistent with discoveries in other fields as well. Unfortunately, the religious fundamentalist can almost never be convinced by what to the scientist is evidence; he puts his faith in religious dicta and denies the evidence of his own senses. That is the crux of the controversy, a matter of fundamental difference in viewpoint—a difference which involves the nature of reality itself. That there is no inconsistency between religious belief and acceptance of evolution is abundantly demonstrated by the great number of biologists and other evolutionists who are active in the support of religious institutions. In fact, the biologist, as others, realizes that no force equals religion as a means of drawing out from man the best qualities which he possesses.

# APPENDIXES

# REFERENCES

Books suggested as references for supplementary reading are here classified in five groups, corresponding to the five parts of this Outline. No textbooks of general biology are included as these are referred to in the Quick Reference Table.

There has recently been an enormous proliferation of brief works devoted to special aspects of biological science. We cannot adequately list the large number of these now available, or becoming so month by month, but many are extremely useful references for the student of biology. Most of them have appeared in paperback, in the following series: *Concepts of Modern Biology, Foundations of Modern Biology, Foundations of Developmental Biology,* and *Foundations of Modern Genetics,* Prentice-Hall; *Contemporary Thought in Biological Science,* Dickenson; *Current Concepts in Biology,* Macmillan; *Frontiers of Biology,* Interscience (Wiley); *Modern Biology,* Holt, Rinehart and Winston; *Principles of Biology,* Addison-Wesley; *Riverside Studies in Biology,* Houghton Mifflin; *Selected Topics in Modern Biology,* Reinhold. Other good references not to be overlooked are reprints of articles that have appeared in *Scientific American,* available singly or in collections from W. H. Freeman.

There are several good series of books on natural history: *Life Nature Library,* Time, Inc.; *Our Living World of Nature,* McGraw-Hill and World Book Encyclopedia; *Pictured-Key Nature Books,* W. C. Brown.

## Part I: Life in Its Simplest Forms

Bonner, James. *The Molecular Biology of Development.* Oxford University Press, 1965.

Borek, Ernest. *The Code of Life.* Columbia University Press, 1965.

Brachet, J. and Mirsky, A. E., eds. *The Cell.* Vols. I-VI. Academic Press, 1959–1964.

Burrows, William. *Textbook of Microbiology.* Saunders, 1968.

Davson, Hugh. *A Textbook of General Physiology.* Little, Brown, 1964.

DeRobertis, Eduardo D., et al. *Cell Biology.* Saunders, 1965.

Downes, Helen R. *The Chemistry of Living Cells.* Harper Bros., 1962.

Fawcett, Don W. *The Cell: Its Organelles and Inclusions.* Saunders, 1966.

Frobisher, Martin. *Fundamentals of Microbiology.* Saunders, 1968.

Giese, Arthur C. *Cell Physiology.* Saunders, 1968.

Haggis, G. H., et al. *Introduction to Molecular Biology.* Wiley, 1964.

International Congress of Zoology, Wash., 1963. *Ideas in Modern Biology.* Ed. by Moore, John A. Natural History Press, 1965.

Kudo, Richard R. *Protozoology*. Thomas, 1966.

Lehninger, Albert L. *Bioenergetics*. Benjamin, 1965.

Oparin, Alexander I. *The Origin of Life*. Trans. by Morgulis, S. Dover, 1953.

Pelczar, Michael J. Jr. and Reid, R. D. *Microbiology*. McGraw-Hill, 1965.

Porter, Keith R. and Bonneville, M. A. *An Introduction to the Fine Structure of Cells and Tissues*. Lea & Febiger, 1968.

Stanier, Roger Y., et al. *The Microbial World*. Prentice-Hall, 1963.

Stern, Herbert and Nanney, David L. *The Biology of Cells*. Wiley, 1965.

Watson, James D. *Molecular Biology of the Gene*. Benjamin, 1965.

Wilson, E. B. *The Cell in Development and Inheritance*. Macmillan, 1928.

## Part II: Multicellular Organisms

Arey, Leslie B. *Developmental Anatomy*. Saunders, 1965.

Buchsbaum, Ralph. *Animals without Backbones*. University of Chicago Press, 1948.

Devlin, Robert M. *Plant Physiology*. Reinhold, 1966.

Elliott, Alfred M. *Zoology*. Appleton-Century-Crofts, 1968.

Esau, Katherine. *Plant Anatomy*. Wiley, 1965.

Florey, Ernest. *An Introduction to General and Comparative Animal Physiology*. Saunders, 1966.

Foster, Adriance S. and Gifford, Ernest M. Jr. *Comparative Morphology of Vascular Plants*. Freeman, 1959.

Greulach, Victor A. and Adams, J. E. *Plants: An Introduction to Modern Botany*. Wiley, 1967.

Holmes, Samuel J. *The Biology of the Frog*. Macmillan, 1927.

Huxley, T. H. *The Crayfish*. D. Appleton, 1906.

Hyman, Libbie H. *The Invertebrates*. 6 vols. (Protozoa through Echinodermata and part of Mollusca.) McGraw-Hill, 1940–1967.

Parker, Thomas J. and Haswell, W. A. *A Textbook of Zoology*. 2 vols. Macmillan, 1962.

Pennak, Robert W. *Fresh-Water Invertebrates of the United States*. Ronald, 1953.

Prosser, C. Ladd and Brown, Frank A. Jr. *Comparative Animal Physiology*. Saunders, 1961.

Romer, Alfred S. *The Vertebrate Body*. Saunders, 1962.

Ross, Herbert H. *A Textbook of Entomology*. Wiley, 1965.

Scheer, Bradley T. *Animal Physiology*. Wiley, 1963.

Smith, Gilbert M. *Cryptogamic Botany*. McGraw-Hill, 1955.

Storer, Tracy I. and Usinger, Robert L. *General Zoology*. McGraw-Hill, 1965.

Swain, Ralph B. *The Insect Guide*. Doubleday, 1948.

Telfer, William and Kennedy, D. *The Biology of Organisms*. Wiley, 1965.

Turner, C. Donnell. *General Endocrinology*. Saunders, 1966.

## Part III: Man: Morphology, Physiology, and Reproduction

Best, Charles H. and Taylor, N. B. *Human Body: Its Anatomy and Physiology*. Holt, Rinehart and Winston, 1963.

———. *The Physiological Basis of Medical Practice.* Williams & Wilkins, 1966.

Frohse, Franz, et al. *Atlas of Human Anatomy.* Barnes & Noble, 1961.

Gray, Henry. *Gray's Anatomy of the Human Body.* Ed. by Goss, C. M. Lea & Febiger, 1966.

Grollman, Sigmund. *The Human Body: Its Structure and Function.* Macmillan, 1969.

Guyton, Arthur C. *Function of the Human Body.* Saunders, 1969.

Kimber, Diana C., et al. *Anatomy and Physiology.* Macmillan, 1966.

Patten, Bradley M. *Human Embryology.* McGraw-Hill, 1968.

Steen, E. B. and Montagu, A. *Anatomy and Physiology.* 2 vols. Barnes & Noble, 1959.

## Part IV: General Principles

Allee, Warder C., et al. *Principles of Animal Ecology.* Saunders, 1949.

Andrewartha, H. G. *Introduction to the Study of Animal Populations.* Methuen, 1961.

Carson, Rachel. *The Sea Around Us.* Oxford University Press, 1961. (Paperback, New American Library.)

Darlington, Philip J. Jr. *Zoogeography: The Geographical Distribution of Animals.* Wiley, 1957.

Daubenmire, Rexford F. *Plants and Environment.* Wiley, 1959.

Dobzhansky, Theodosius G. *Genetics and the Origin of Species.* Columbia University Press, 1951.

Dodson, Edward O. *Evolution: Process and Product.* Reinhold, 1960.

Herskowitz, Irwin H. *Genetics.* Little, Brown, 1965.

Hesse, Richard, et al. *Ecological Animal Geography.* Wiley, 1951.

Howells, William W. *Mankind in the Making.* Doubleday, 1967.

Hough, John N. *Scientific Terminology.* Rinehart, 1953.

Kendeigh, Samuel C. *Animal Ecology.* Prentice-Hall, 1961.

King, Robert C. *Genetics.* Oxford University Press, 1965.

Lorenz, Konrad. *King Solomon's Ring.* Crowell, 1952. (Paperback, Apollo.)
———. *On Aggression.* Harcourt, Brace, 1966. (Paperback, Bantam.)

MacArthur, Robert H. and Connell, J. H. *The Biology of Populations.* Wiley, 1966.

Marler, Peter and Hamilton, W. J. *Mechanisms of Animal Behavior.* Wiley, 1966.

Mayr, Ernst. *Animal Species and Evolution.* Harvard University Press, 1963.

———. *Principles of Systematic Zoology.* McGraw-Hill, 1969.

Moody, Paul A. *Genetics of Man.* Norton, 1967.

Odum, Eugene P. and Odum, Howard T. *Fundamentals of Ecology.* Saunders, 1959.

Oosting, Henry J. *The Study of Plant Communities.* Freeman, 1956.

Orr, Robert T. *Vertebrate Biology.* Saunders, 1966.

Ross, Herbert H. *A Synthesis of Evolutionary Theory.* Prentice-Hall, 1962.

Savory, Theodore. *Naming the Living World.* Wiley, 1963.

Scott, John P. *Animal Behavior.* University of Chicago Press, 1958.

Simpson, George G. *The Meaning of Evolution.* Yale University Press, 1967.

——— et al. *Quantitative Zoology.* Harcourt, Brace, 1960.

Srb, M. A., et al. *General Genetics.* Freeman, 1965.

Stern, Curt. *Principles of Human Genetics.* Freeman, 1960.

Winchester, A. M. *Heredity.* Barnes & Noble, 1966.

———. *Genetics.* Houghton Mifflin, 1966.

Woods, R. S. *The Naturalist's Lexicon.* Abbey Garden Press, 1944; and *Addenda,* 1947.

## Part V: Relations of Biology to Mankind

Anderson, Edgar. *Plants, Man and Life.* University of California Press, 1952.

Beckner, Morton. *The Biological Way of Thought.* University of California Press, 1968.

Carson, Rachel. *Silent Spring.* Houghton Mifflin, 1962. (Paperback, Fawcett World.)

Commoner, Barry. *Science and Survival.* Viking, 1963.

Conant, James B. *Modern Science and Modern Man.* Columbia University Press, 1952. (Paperback, Doubleday, 1953.)

Dasmann, Raymond F. *Environmental Conservation.* Wiley, 1968.

Dobzhansky, Theodosius G. *Heredity and the Nature of Man.* Harcourt, Brace & World, 1964. (Paperback, New American Library.)

Dubos, Rene. *The Torch of Life.* Simon & Schuster, 1962. (Paperback, Pocket Books.)

Dunn, L. C. and Dobzhansky, Th. *Heredity, Race and Society.* New American Library, 1952.

Farb, Peter. *Face of North America: The Natural History of a Continent.* Harper & Row, 1963.

Hall, Thomas S. *A Source Book in Animal Biology.* McGraw-Hill, 1951.

Henderson, Lawrence J. *The Fitness of the Environment.* Peter Smith, 1959.

Huxley, Julian. *Man in the Modern World.* New American Library, 1948.

Montagu, Ashley. *Man's Most Dangerous Myth: The Fallacy of Race.* World, 1964.

Peattie, Donald C. *Green Laurels.* Literary Guild, 1936.

Robbins, W. W. and Ramaley, F. *Plants Useful to Man.* Blakiston, 1933.

Singer, Charles. *A History of Biology.* Abelard, 1959.

Sinnott, Edmund W. *Cell and Psyche.* University of North Carolina Press, 1950.

Sirks, Marius J. and Zirkle, Conway. *The Evolution of Biology.* Ronald, 1964.

Udall, Stewart L. *The Quiet Crisis.* Holt, Rinehart and Winston, 1963.

# AN ABRIDGED CLASSIFICATION OF LIVING ORGANISMS

Most current textbooks of biology classify all organisms in either the plant kingdom or the animal kingdom. Several recognize a special group, the Kingdom Monera, which includes all organisms consisting of procaryotic cells. (If only two kingdoms are recognized such organisms are included in the plant kingdom.) A few textbooks include a Kingdom Protista, but its composition varies with different authors. The various treatments of major categories of classification include: recognition of two kingdoms only, Plantae and Animalia; recognition of three kingdoms, which may be either Monera, Plantae, and Animalia, or Protista, Plantae, and Animalia; recognition of four kingdoms, Monera, Protista, Plantae, and Animalia. In the following condensed summary we recognize three kingdoms, Monera, Plantae, and Animalia:

KINGDOM MONERA
  Phylum Schizophyta. Bacteria.
  Phylum Cyanophyta. Blue-green algae.
KINGDOM PLANTAE
  Subkingdom Thallophyta.
  Phylum Eumycophyta. True fungi.
  Phylum Myxomycophyta. Slime molds.
  Phylum Chlorophyta. Green algae.
  Phylum Chrysophyta. Diatoms.
  Phylum Pyrrophyta. Dinoflagellates.
  Phylum Euglenophyta. Euglenoid flagellates.
  Phylum Phaeophyta. Brown algae.
  Phylum Rhodophyta. Red algae.
  Subkingdom Embryophyta.
  Phylum Bryophyta.
    Class Musci. Mosses.
    Class Hepaticae. Liverworts.
    Class Anthocerotae. Hornworts.
  Phylum Tracheophyta.
    Subphylum Psilopsida. Psilopsids.
    Subphylum Lycopsida. Club mosses.
    Subphylum Sphenopsida. Horsetails and their relatives.
    Subphylum Pteropsida. Ferns and seed plants.

Class Filicineae. Ferns.
Class Gymnospermae. Cone-bearing plants and their relatives.
Class Angiospermae. True flowering plants.
  Subclass Dicotyledoneae.
  Subclass Monocotyledoneae.

# KINGDOM ANIMALIA
Phylum Protozoa. Unicellular animals.
    Class Mastigophora (Flagellata). Flagellate protozoa.
    Class Sarcodina (Rhizopoda). Amoeboid protozoa.
    Class Sporozoa. Spore-producing protozoa.
    Class Ciliata (Infusoria). Ciliate protozoa.
Phylum Porifera. Sponges.
Phylum Coelenterata (Cnidaria).
    Class Hydrozoa. *Hydra* and other hydroids.
    Class Scyphozoa. Jellyfishes.
    Class Anthozoa. Corals and sea anemones.
Phylum Ctenophora. Comb jellies or sea walnuts.
Phylum Platyhelminthes. Flatworms.
    Class Turbellaria. Free-living flatworms.
    Class Trematoda. Flukes.
    Class Cestoda. Tapeworms.
Phylum Nemertea. Nemertine worms.
Phylum Aschelminthes.
    Class (or Phylum) Nematoda. Roundworms.
    Class (or Phylum) Rotifera. Rotifers.
    Class (or Phylum) Nematomorpha. Horsehair worms.
Phylum Acanthocephala. Spiny-headed worms.
Phylum Ectoprocta. Bryozoa.
Phylum Brachiopoda. Lamp shells.
Phylum Mollusca.
    Class Amphineura. Chitons.
    Class Gastropoda. Snails, slugs.
    Class Scaphopoda. Tooth shells.
    Class Pelecypoda. Clams, mussels, oysters.
    Class Cephalopoda. Squids, octopus, nautilus.
Phylum Annelida. Segmented worms.
    Class Chaetopoda. Earthworms, clamworms.
    Class Hirudinea. Leeches.
Phylum Onychophora. *Peripatus*.
Phylum Arthropoda.
    Class Crustacea. Crayfish, crabs, barnacles, water fleas.
    Class Diplopoda. Millipedes.
    Class Chilopoda. Centipedes.
    Class Insecta. Insects: grasshoppers, bees, beetles, flies, etc.
    Class Arachnida. Ticks, spiders, scorpions, horseshoe crab.
Phylum Echinodermata.
    Class Asteroidea. Starfishes.
    Class Ophiuroidea. Brittle stars, serpent stars.

Class Echinoidea. Sea urchins, sand dollars.
Class Holothuroidea. Sea cucumbers, sea slugs.
Class Crinoidea. Sea lilies.
Phylum Hemichordata. Acorn worms and relatives.
Phylum Chordata
  Subphylum Urochordata. Tunicates: sea squirt, sea pork.
  Subphylum Cephalochordata. *Amphioxus.*
  Subphylum Vertebrata.
    Class Agnatha. Lampreys, hagfishes.
    Class Chondrichthyes. Cartilaginous fishes: sharks, rays.
    Class Osteichthyes. Bony fishes: perch, trout, catfish, etc.
    Class Amphibia. Salamanders, frogs, toads.
    Class Reptilia. Turtles, snakes, lizards, alligators, dinosaurs.
    Class Aves. Birds: ostrich, chicken, owl, sparrow, etc.
    Class Mammalia. Mammals: opossum, rabbit, bat, whale, horse, man, etc.

# Appendix C

# SELECTED METHODS OF
# BIOLOGICAL INVESTIGATION

This appendix summarizes some representative biological techniques used in the study of protoplasm and the cell. Biology depends upon techniques developed through physics and chemistry as well as those relating directly to living systems. Some techniques, such as the use of the light and electron microscopes, extend our analysis to levels of organization too small to be seen with the unaided eye. Other techniques, such as electrophoresis and selective staining, use specific properties of compounds found in living systems as a means of identifying and separating them. Still other techniques, such as the use of radioactive tracers to label compounds, enable biologists to follow the functional fate of specific molecules.

Our discussion is divided into two sections: (1) the study of protoplasm and the chemical nature of biological compounds, and (2) the study of cells. Some areas of research techniques, such as statistics and environmental research methods, are not represented in this discussion.

## THE STUDY OF PROTOPLASM

Methods of biological analysis of protoplasm involve a large variety of relatively standard, although often modified, laboratory techniques. The environment of the cell or organism and the protoplasm itself can be measured for the concentration or level of various inorganic materials and factors. Many of these measurements involve the use of meters that have been modified for special purposes. Examples are thermometers, manometers, meters coupled with pH or other electrodes, and flame photometers. The organic constituents of the cell can be separated, identified, and otherwise tested through the use of various chromatographic and spectrophotometric methods.

Because respiration and photosynthesis involve exchanges of oxygen and carbon dioxide, variations in the concentration of these gases reflect the rates of the reaction sequences. The rate of respiration can be determined by absorbing the released carbon dioxide and measuring the corresponding decrease in air volume. The rate of photosynthesis can be determined by supplying a source of carbon dioxide and measuring the volume of oxygen given off. In both cases the observations are carried out under uniform temperature and pressure.

In *paper chromatography* a drop of solution containing the substances to be separated is applied toward the end of a filter paper strip. The paper tip at this end is then placed in a solvent, which travels up the paper and moves various components of the mixture away from the spot at different rates, these rates being dependent upon molecular size, adsorption to the paper, and other molecular characteristics. The different organic compounds generally form easily detected bands. Various treatment techniques color or otherwise identify and separate the organic compounds that could not have been easily detected in the mixed form. Modifications of this technique involve the use of more than one solvent and the use of paper at different angles and of different shapes (e.g., circular development). In *two-dimensional chromatography,* commonly used for mixtures of many amino acids, the paper is rotated at right angles to the first separation and a second solvent separation is made. *Thin-layer chromatography* uses thin layers of a variety of absorbent media other than paper, uniformly applied. *Column chromatography* is the separation of compounds by passing the mixture through columns of porous material; the compounds are adsorbed differentially and can then be isolated. In *gas chromatography* the moving liquid solvent that carries the mixture is replaced by a gas. This technique is often more rapid and more accurate than use of liquids, and it can be conducted over a wider range of temperature. In *electrophoresis,* the separation involves migration of organic compounds in an electric field. The direction and rate of migration are influenced by the ionic strength and buffer concentration in the solution, as well as by molecular differences in shape and electric charge.

Another method of detecting organic compounds (often in mixture form) is *absorption spectroscopy.* This technique is based on the fact that certain kinds of chemicals and chemical bonds absorb specific wavelengths of light; thus, if a beam of light is passed through a chemical solution the absorption reduces and selectively alters the radiation coming out of the solution. The instruments used for this analysis include color comparators, colorimeters, and spectrophotometers for ultraviolet, visible, and infrared absorption analysis. Qualitative analysis of an unknown chemical may be made by determining its absorption spectrum. In infrared spectroscopy the absorption bands are controlled by atomic mass and bond characteristics; the physical data produced can be used in making specific molecular identifications. Because in many cases the light absorbed is proportional to the concentration of the chemical under investigation, quantitative analysis is often possible. Thus molecular identity, structure, and concentration can be determined through knowledge of absorption. These techniques are valuable in biological studies because of their speed and reliability, and because they can be used to study mixtures, low concentrations, and chemicals in intact cells and tissues, allowing the investigation to follow reaction sequences without disrupting the cell architecture.

The use of *isotopes,* especially *radioactive isotopes,* to construct and thereby label specific compounds, has provided a basis for the functional study of metabolic transformations and the environmental fate of specific compounds and their products. Radiation emitted by the isotopes is readily detected by a Geiger-Müller counter. In *autoradiography*, a cell or tissue preparation that has been cultured in a medium containing radioactive isotopes is exposed to a special photographic film. The radiation from the isotopes affects the film adjacent to their positions in the tissue. This process has aided investigators in finding the specific location of numerous biological activities.

## CELL STUDY: PREPARATION AND MICROSCOPY

Methods of study of cells require various techniques of preparation appropriate to the types of optical instruments used in their examination. These techniques are primarily methods of staining (developed to produce contrasts between cell components) and the mechanical preparation of cells and tissues to meet requirements of the types of microscopes used.

The *compound microscope* ordinarily magnifies an image of the object studied from about one hundred to one thousand diameters. This makes it possible visually to observe cells and much of their contents. The compound microscope has two sets of lenses: the objective lens system, which produces an enlarged real image, and the ocular or eyepiece lenses, which produce a virtual image at the distance where the human eye is adapted to see objects most clearly. Although the compound microscope generally uses light transmitted through the object observed, the *dissecting microscope,* of lower magnifications, may use reflected light.

There are numerous compound microscope modifications, involving both the light source and the lens system, that increase contrast between cell components and obtain greater resolution of cell details. The *phase-contrast microscope* uses two sets of light rays to exaggerate differences in the velocity of light penetration of adjacent structures. This, for example, enables the viewer to follow the activity of chromosomes in living cells during mitosis or meiosis. The *dark-field microscope* has a special light condenser that directs the light from the center of the field; light scattered by refraction reveals structural edges and small objects, which appear bright against a dark background. The *polarizing microscope* has optics that transmit and analyze polarized light. Polarizing optics detect birefringence, which is exhibited by objects that refract light in two different directions. Birefringence in biological materials indicates that the molecules have a regular or crystallinelike orientation. The polarizing microscope is used to study fibrillar structures, membranes, and cell division.

The use of television and camera attachments is a helpful adjunct to various aspects of microscopy.

Temporary preparations for examination with the light microscope can often be made with *vital dyes,* stains that give color to substances in living cells without causing immediate serious injury. These dyes are used to stain subcellular structures selectively or to mark specific cells so that movements during embryonic development can be followed.

Permanent *slide preparations* for use with a light microscope involve the following steps: (1) The cells are *killed* as quickly as possible, to avoid change. The agents used are selected for great toxicity and power of rapid penetration. (2) The protoplasm is *fixed* or coagulated in as nearly normal condition as possible. Often the same reagent serves both as a killing and fixing agent (e.g., formalin, alcohol). Many errors in observation are due to the use of inappropriate killing and fixing agents. (3) Water is very gradually removed from the preparation (*dehydration*) and replaced by another liquid. Alcohols of increasing concentration are most commonly used as dehydrating agents. (4) After being impregnated with and *imbedded* in paraffin or celloidin, the preparation is *sectioned* by a *microtome*. The microtome cuts thin, regular sections from five to twenty microns in thickness (a micron being 1/1,000 millimeter, or about 1/25,000 inch). (5) Ordinarily one or more times during the preparation process the cells are *stained* with dyes of known chemical properties. Basic dyes (e.g., basic fuchsin) stain the nucleus, and acid dyes (e.g., eosin) stain the cytoplasm. (6) The thin sections are *cleared* of paraffin or celloidin through the use of a clearing agent (e.g., xylol) and *mounted* under thin glass slips (cover glasses) on microscope slides. The mounting medium is often a gum that hardens on standing, forming a clear substance through which the cell and its stained contents can be observed.

The *transmission electron microscope* uses a stream of electrons instead of light. Focused by electromagnets, the stream of electrons passes through the object under investigation to a fluorescent screen or a photographic plate. The electrons are scattered by the object, the scattering being greater the denser the material, regardless of its chemical composition. If the object is too thick all the electrons will be scattered. Thus, special *ultramicrotomes* have been developed to cut exceptionally thin sections (0.2 to 0.002 micron). The electron microscope can produce a magnification of over one hundred thousand diameters. Various methods are used to exaggerate differences within the cells studied, either treating the biological compounds with electron-dense material (called positive staining) or filling the interstices with heavy metal compounds (called negative staining). *Shadowcasting*, the vacuum evaporation of heavy metals at oblique angles to the specimen, may be carried out to outline objects (by the metal accumulation) that otherwise do not contrast with their background. Because gas also scatters electrons the process must be performed in a vacuum, and the preparation must be dry to eliminate water vapor. Several techniques have been developed to prepare material for electron microscope study, including the skimming of materials concentrated on surface films, construction of casts of the objects desired to be studied, and freeze-drying and fracturing preparations. Studies with the electron microscope have contributed much to our knowledge of cellular and subcellular structures.

The use of high-speed centrifuges (*ultracentrifugation*) to separate cell components has been important in preparing material for the electron microscope and for physiological studies of the functional differences between organelles. The cellular material to be centrifuged is often first subjected to mechanical disintegration within specific solvents, after which progressively smaller components of the cells are separated at increasing speeds.

In the *scanning electron microscope* a primary electron beam scanned across a specimen (which may be thinly coated for grounding purposes) causes the emission of electrons, which can be used to produce a three-dimensional image. The ability to use solid specimens, the wide range in magnification (50 to 25,000 diameters), and the large depth of focus make this a very useful instrument for the study of intact single-celled and multicellular organisms as well as for the study of cell structure.

# GLOSSARY

The definitions given below are brief, even terse. It is intended that they suggest the ordinary usage of each term in biology, and that they serve primarily to identify, rather than explain completely, the terms given. Less common terms, defined where they appear in the Outline, are not included, nor are names of taxonomic entities; see the index for these.

**Abdomen:** the major body division posterior to head and thorax.

**Actin:** *See* Actomyosin.

**Active transport:** movement of materials through the cell membrane against a diffusion gradient.

**Actomyosin:** combination of actin and myosin, muscle proteins, responsible for contraction.

**ADP:** adenosine diphosphate; nucleotide involved in energy mobilization.

**Aerobic:** carrying on respiration in the presence of free oxygen.

**Agonistic behavior:** aggressive behavior and behavior in response to aggression.

**Alimentary canal:** the digestive tract, from mouth to anus.

**Allantois:** extraembryonic membrane that carries respiratory blood vessels of embryos of higher vertebrates.

**Allele:** one of two or more alternative genes.

**Alternation of generations:** alternation of gametophyte and sporophyte generations in plants.

**Amino acid:** an organic acid with an amino ($-NH_2$) group; a building block of proteins.

**Amitosis:** nuclear division without formation of a spindle.

**Amnion:** extraembryonic membrane immediately surrounding the embryos of higher vertebrates.

**Anabolism:** constructive metabolism; metabolism involving energy storage.

**Anaerobic:** carrying on respiration in the absence of free oxygen.

**Analogy:** superficial similarity, due only to similarity in function. *See* Homology.

**Anatomy:** the study of visible or gross structure.

**Androecium:** the stamens collectively.

**Anterior:** toward the forward end.

**Anther:** the pollen sacs of the stamen.

**Antheridium:** a plant organ in which male gametes are formed.

**Antibody:** a protein produced in the body to combat the injurious effect of a foreign substance (antigen).

**Antigen:** a foreign protein that causes the production of an antibody in the organism.

**Anus:** outlet of the digestive tract.

**Archegonium:** a plant organ in which female gametes are formed.

**Artery:** a blood vessel that carries blood away from the heart.

**Asexual reproduction:** reproduction by one individual, independent of others.

**Assimilation:** the manufacture of reserve food or protoplasm.

**ATP:** adenosine triphosphate; nucleotide involved in energy mobilization.

**Atrium:** a chamber of the heart in which blood is received.

**Auricle** = atrium.

**Autecology:** the ecology of an individual organism or species.

**Autonomic nervous system:** that portion of the nervous system independent of voluntary control.

**Autotrophic:** a type of nutrition in which the organism manufacturers its own food.

**Axon:** any single-fiber extension, "conductile region," of a neuron; formerly considered a fiber carrying impulses away from the cell body.

**Auxin:** a plant hormone; it stimulates cell elongation and has other functions.

**Biogenetic law:** the statement that the development of the individual repeats the development of the race.

**Biology:** the science of life.

**Biome:** a major biotic community characterized by a predominant life form.

**Biotic potential:** an expression of the rate of reproduction of an organism.

**Blastocyst:** blastula stage of a mammalian embryo.

**Blastopore:** the opening in the gastrula of an animal embryo.

**Blastula:** an early stage of the embryo, which consists essentially of a hollow ball of cells.

**Botany:** the science of plant life.

**Calyx:** the outer whorl of a complete flower; the sepals collectively.

**Cambium:** meristematic tissue responsible for secondary thickening in stem or root.

**Cancer:** abnormal, rapidly growing and dividing tissue.

**Capillary:** a small thin-walled blood vessel connecting an artery with a vein or (in portal systems) a vein with a vein.

**Carbohydrate:** a sugar or a condensation product of sugars.

**Carpel:** a megasporophyll of a flowering plant; one of the units of the gynoecium.

**Catabolism:** destructive metabolism; metabolism involving release of energy.

**Cell:** the unit of structure and function in organisms.

**Central body** = centrosome.

**Centromere:** place of attachment of a mitotic fiber on a chromosome.

**Centrosome:** an organelle, present in cells of animals and lower plants, from which radiate spindle fibers during mitosis.

**Cephalothorax:** the head and thorax combined in one structure.

**Chemosynthesis:** synthesis in which the energy for the synthesis is derived from a chemical reaction.

**Chlorophyll:** the pigment of green plants involved in photosynthesis.

**Chloroplast:** a plastid containing chlorophyll.

**Chorion:** outermost embryonic membrane in higher vertebrates, involved in formation of placenta in mammals.

**Chromatid:** one of the two strands constituting a chromosome prior to its division.

**Chromatin:** a nuclear constituent staining readily with basic dyes; it consists of DNA and protein.

**Chromosome:** a structure formed of chromatin that appears in cells during mitosis; bearer of DNA and protein.

**Citric acid cycle:** the cycle of chemical reactions in aerobic respiration by which the carbons of the acetyl group are oxidized to carbon dioxide; also called Krebs cycle.

**Classification:** the grouping of organisms on the basis of fundamental similarity.

**Cleavage stages:** the early stages of cell division in an animal embryo.

**Climax community:** a self-perpetuating community characteristic of a given climate; the final stage in an ecological succession.

**Cloaca:** the common terminal channel for digestive, excretory, and reproductive systems in vertebrates; the terminal portion of the digestive tract in certain invertebrates.

**CoA:** coenzyme A, a compound involved in the transfer of the acetyl group to the citric acid cycle in the aerobic phase of cell respiration.

**Codon:** a sequence of three nucleotides in messenger RNA.

**Coelom:** the body cavity; the space, lined with mesoderm, in which the viscera lie.

**Colony:** a group of organisms of the same species living together.

**Community:** a group of organisms interrelated by environment requirements.

**Conjugation:** the union of two unicellular organisms prior to sexual reproduction.

**Corolla:** next to outer whorl of complete flower; the petals collectively.

**Corpuscle:** a blood cell.

**Cotyledon:** an embryo leaf.

**Cranium:** that portion of the skull that surrounds the brain.

**Crossing over:** the exchange of genetic material between homologous chromosomes during meiosis.

**Cytochrome:** a compound involved in hydrogen transport during cell respiration.

**Cytokinesis:** the division of the cell body.

**Cytology:** the study of cells.

**Cytoplasm:** the protoplasm of the cytosome.

**Cytosome:** that part of the cell outside the nucleus.

**Dendrite:** the branching portion or generator region of a neuron; formerly considered the portion of a neuron conducting an impulse toward the cell body.

**Deuterostome:** an animal in which the mouth develops from a new embryonic opening not near the blastopore.

**Differentiation:** the process by which different types of cells, tissues, or organs are derived from a common pattern.

**Digestion:** preparation of food for absorption and assimilation, by hydrolysis.

**Dioecious:** condition in which male and female flowers occur on different plants; condition in animals in which the male and female gametes are produced in different individuals.

**Diploid:** the chromosome number in which the chromosomes are represented by homologous pairs; twice the haploid number; the $2n$ number.

**Distal:** away from the point of attachment or place of reference.

**Division of labor:** functional differentiation.

**DNA:** deoxyribonucleic acid, carrier of genetic information.

**Dominance:** the condition in which one of two alternative characters is evident in a heterozygote (genetics); prominence in a biotic community (ecology); the condition in which one organism occupies a position of control over another (animal behavior).

**Dorsal:** toward the back.


**Ecology:** the study of relations between organism and environment.

**Ecosystem:** a biotic community and its abiotic environment considered as a unit.

**Ecotone:** the region of overlap between two biotic communities.

**Ectoderm:** outer layer of cells of the early embryo.

**Egestion:** the discharge of unabsorbed food from an animal.

**Egg cell:** a female gamete.

**Embryo:** an early stage in the development of an organism.

**Embryology:** the study of development.

**Embryo sac:** female gametophyte of higher plants.

**Endocrine gland:** a gland of internal secretion (ductless gland); a gland that produces a hormone.

**Endoderm:** innermost layer of cells in the gastrula stage of the embryo.

**Endoplasmic reticulum:** double-membrane network in the cytoplasm.

**Endoskeleton:** internal supporting structure.

**Endosperm:** triploid nutrient tissue in seeds of angiosperms.

**Enterocoel:** a coelom formed by fusion of the cavities of mesodermal pouches in the embryo.

**Enzyme:** a catalyst characteristic of living organisms.

**Epidermis:** a layer of cells covering an external surface.

**Epithelium:** a layer of cells covering a surface or lining a cavity in animals.

**Erythrocyte:** a red blood cell, one that contains hemoglobin.

**Eucaryotic cell:** a cell with a definite nucleus and other double-membrane organelles.

**Eugenics:** the improvement of man by improvement of his heredity.

**Euthenics:** the improvement of man by improvement of his environment.

**Evolution:** the process of descent with change.

**Excretion:** the discharge of waste materials formed in metabolism.

**Exoskeleton:** external supporting structure.

**Fat:** a glyceryl ester of fatty acids; compound of glycerol and three fatty acid molecules.

**Fermentation:** anaerobic respiration resulting in the production of alcohol or another incompletely oxidized end product.

**Fertilization:** the union of two gametes to form a zygote.

**Fetus:** the human embryo; term usually applied after about the third month of pregnancy.

**Fission:** asexual reproduction by division into two equivalent parts.

**Flower:** the organ in angiosperms in which the sexually reproductive structures are contained.

**Food chain:** a series of organisms that represent the sequence of food from producers to ultimate consumers.

**Free-living:** independent, as applied to an organism; opposite of parasite.

**Fruit:** a ripened ovary; the seed-containing structure of a flowering plant.

**Gamete:** a mature germ cell; sperm or egg.

**Gametogenesis:** the process of gamete formation.

**Gametophyte:** gamete-producing generation in plants.

**Ganglion:** a concentration of nerve cells.

**Gastrodermis:** layer of cells lining the gastrovascular cavity.

**Gastrovascular cavity:** a combined digestive and circulatory cavity.

**Gastrula:** an early stage of the embryo that consists essentially of an invaginated blastula.

**Gene:** an hereditary determiner, located in a chromosome; part of a DNA molecule.

**Genetics:** the science of heredity.

**Genotype:** the fundamental hereditary (genetic) constitution of an organism. *See* Phenotype.

**Germ layer:** one of the three embryonic cell layers of multicellular animals.

**Germination:** the process by which the embryo in a seed breaks the seed coat and begins to develop into a young plant.

**Germ plasm:** the gametes and the cells from which they are formed.

**Genetic code:** the morphological basis of inheritance as determined by sequences in messenger RNA.

**Glycogen:** a complex polysaccharide; "animal starch."

**Glycolysis:** the anaerobic phase of cellular respiration; the breakdown of glucose to pyruvic acid.

**Golgi apparatus:** a membranous organelle of the cytoplasm, presumably of secretory function.

**Gonad:** a gamete-producing organ in animals; testis or ovary.

**Growth:** increase in size.

**Gynoecium:** the carpels collectively.

**Habitat:** the specific place where an organism lives; a region characterized by particular environmental factors.

**Haemocoel:** a body cavity that functions as a part of the blood-vascular system.

**Haploid:** the chromosome number representing a single set of chromosomes; the *n* number.

**Hemoglobin:** the oxygen-carrying protein of the blood.

**Heredity:** transmission of characters from parent to offspring.

**Hermaphroditism:** the condition in which gonads of both sexes occur in the same individual.

**Heterogamy:** the condition in which the gametes of opposite sex are morphologically distinct.

**Heterotrophic:** a type of nutrition in which the organism gets its food from other organisms.

**Heterozygous:** said of an individual that has unlike or alternative genes (allelomorphs) for the character being considered. *See* Homozygous.

**Histochemistry:** the chemistry of cells and tissues.

**Histology:** the study of tissues.

**Homology:** fundamental similarity, based primarily on development and structure. *See* Analogy.

**Homozygous:** said of an individual that has duplicate genes for the character being considered. *See* Heterozygous.

**Hormone:** a chemical regulator; a substance that serves in the chemical coordination of the body.

**Hyphae:** filaments characteristic of the fungi.

**Immunity:** resistance to disease.

**Independent assortment:** independent inheritance of characters whose determiners are on different pairs of chromosomes. (Mendel's second law.)

**Induction:** the influence of a group of cells in modifying the pattern of differentiation in other, associated cells.

**Integument:** the outer covering of the body.

**Irritability:** the capacity of responding to a stimulus.

**Isogamy:** the condition in which the gametes of opposite sex are morphologically alike.

**Isozyme:** a variant form of an enzyme.

**Karyokinesis:** nuclear division.

**Krebs cycle** = citric acid cycle.

**Leaf:** a generally broad, flat organ of plants, bearing photosynthetic cells.

**Leucocyte:** a white blood cell.

**Linkage:** the condition in which characters are inherited together because their genes are present in the same chromosome.

**Lipid:** any fatlike compound.

**Locus:** the site on a chromosome where a gene occurs.

**Lymphatic system:** the portion of the vascular system in animals that returns intercellular fluid to the veins.

**Lysosome:** a cytoplasmic organelle associated with intracellular digestion.

**Mantle:** the membrane lining the respiratory cavity of mollusks.

**Medusa:** the jellyfish stage of a coelenterate.

**Megagametophyte:** female gametophyte; embryo sac.

**Megaspore:** the larger of the two types of spores in heterosporous plants; it forms the female gametophyte; ovule.

**Megasporophyll:** an organ producing megaspores. In a strobilus, a scale; in a flower, a carpel.

**Meiosis:** cell division in which the chromosome number is reduced from diploid to haploid; it involves two successive cell divisions.

**Mendelism:** the principles of heredity discovered by Gregor Mendel; in particular, segregation and independent assortment.

**Menstrual cycle:** the mammalian female sex cycle, which, in man, is characterized by 28-day intervals between successive losses of uterine lining tissue.

**Meristem:** plant tissue consisting of actively growing and dividing cells.

**Mesoderm:** the layer of cells that forms between ectoderm and endoderm in animal embryos.

**Mesoglea:** the jellylike layer between epidermis and gastrodermis in the Coelenterata and Ctenophora.

**Metabolism:** the chemical processes characteristic of protoplasm.

**Metagenesis:** alternation of sexual and asexual methods of reproduction.

**Metamerism:** segmental arrangement of organs or organ systems.

**Metamorphosis:** pronounced change in form during the course of development.

**Microgametophyte:** male gametophyte; pollen tube.

**Microspore:** pollen grain; the smaller of the two types of spores in heterosporous plants; forms the male gametophyte.

**Microsporophyll:** an organ producing microspores (pollen grains). In a strobilus, a scale; in a flower, an anther.

**Mitochondrion:** cytoplasmic organelle involved in cellular respiration.

**Mitosis:** nuclear division in which the chromosomes are divided equally.

**Molecular biology:** the biochemistry of the large molecules associated with living organisms.

**Monoecious:** condition in which male and female flowers are separate but borne on the same plant.

**Morphology:** the study of structure.

**Mutation:** an inherited change due to a change in a gene.

**Mycelium:** the hyphae of a fungus, considered together.

**Myofibril:** elongated contractile structure within a muscle cell.

**Myosin:** *See* Actomyosin.

**NAD:** nicotinamide adenine dinucleotide: compound involved in hydrogen transport in cellular respiration.

**NADP:** nicotinamide adenine dinucleotide phosphate: compound involved in hydrogen transport in cellular respiration.

**Natural selection:** the process leading to survival and reproduction of those best adapted (most fit) in a natural population.

**Nephridium:** an organ of excretion; a kidney unit.

**Nerve impulse:** the wave of negativity transmitted along a neuron.

**Neuron:** a nerve cell.

**Nomenclature:** the naming of biological species.

**Notochord:** a gelatinous, stiffening, axial support characteristic of all chordates.

**Nucleic acid:** an organic compound made up of nucleotides.

**Nucleolus:** an intranuclear organelle containing RNA.

**Nucleotide:** one of the units of a nucleic acid molecule, consisting of a nitrogen base, a pentose sugar, and a phosphate group.

**Nucleus:** a definite body within a cell, containing chromatin and surrounded by a double membrane.

**Nutrient:** an inorganic constituent essential for plant growth.


**Ocellus:** a simple eye as present in some invertebrates.

**Ommatidium:** a single optical unit in a compound eye.

**Ontogeny:** the course of embryonic development.

**Oögenesis:** the process of maturation of egg cells.

**Organ:** a group of cells or tissues functioning as a unit.

**Organelle:** an intracellular structure.

**Organism:** an individual living thing.

**Organizer:** a region in an embryo that influences the nature of differentiation in adjacent tissues.

**Osmosis:** diffusion of a solvent through a membrane not permeable to the solute, with movement of solvent from region of low to region of high concentration of the solute.

**Ovary:** in plants, the swollen portion of the pistil, in which the ovules develop; in animals, the female gonad, producing eggs.

**Oviduct:** the tube by which the eggs leave the body of the female animal.

**Ovule:** the structure in seed plants that develops into the seed; it is borne on a megasporophyll.

**Ovum:** an egg; a female gamete.


**Paleontology:** the study of fossils.

**Parasite:** an organism that lives at the expense of another.

**Parenchyma:** plant tissue consisting of rounded, thin-walled cells.

**Parthenogenesis:** reproduction by development from an unfertilized egg.

**Pathogenic:** disease producing.

**Pectoral appendage:** one of the anterior pair of paired appendages in vertebrates.

**Pelvic appendage:** one of the posterior pair of paired appendages in vertebrates.

**Peptide:** a compound formed of two or more linked amino acids.

**Perianth:** the calyx and corolla collectively.

**Pericardium:** a cavity surrounding the heart; also, the membrane covering the heart and lining the pericardial cavity.

**Pericycle:** a region in stem or root enclosing the vascular tissues of the plant.

**Peristalsis:** rhythmic muscular contractions that pass along a tubular organ.

**Peritoneal cavity:** the abdominal cavity.

**Petal:** one of the units of the corolla of a flower.

**pH:** the logarithm of the reciprocal of the $H+$ ion concentration.

**Phagocytosis:** movement of solids into a cell by infolding of the cell membrane.

**Pharynx:** the portion of the alimentary tract (and, in vertebrates, the respiratory tract) immediately back of the mouth.

**Phenotype:** the appearance of an organism, without regard to its hereditary constitution. *See* Genotype.

**Phloem:** vascular tissue involved chiefly in a transfer of food materials within the plant.

**Phosphorylation:** addition of a phosphate radical to a molecule.

**Photoperiod:** day length; period of time something is exposed to light.

**Photosynthesis:** synthesis with energy from light; specifically, the synthesis of carbohydrates by green plants in the presence of sunlight.

**Phylogeny:** the course of evolutionary development.

**Phylum:** a major group of animals or plants.

**Physiology:** the study of function.

**Pistil:** a structural unit of the flower, formed of one carpel or of several united.

**Placenta:** structure of maternal and fetal tissues formed in the wall of the mammalian uterus; structure through which diffusion occurs between maternal and fetal blood.

**Plasma:** the liquid portion of the blood.

**Plastid:** a specialized cytoplasmic body.

**Pleural cavity:** the cavity in which the lungs of mammals lie.

**Pollen grain** = microspore.

**Pollen tube** = microgametophyte or male gametophyte.

**Polyp:** the hydroid stage of a coelenterate.

**Polypeptide:** a compound formed of numerous linked amino acids.

**Polysaccharide:** a large molecule formed by condensation of many molecules of sugar.

**Population:** a group of individuals of the same species.

**Posterior:** toward the hinder end.

**Priority:** the taxonomic principle by which the first scientific name given a species is considered the valid one.

**Procaryotic cell:** a cell lacking a definite nucleus and other double-membrane organelles.

**Protein:** a large organic compound made up of amino acids.

**Protoplasm:** living matter.

**Protostome:** an animal in which the mouth develops at or near the blastopore of the embryo.

**Proximal:** near the point of attachment or place of reference.

**Pseudocoelom:** a body cavity between digestive tract and epidermis that is not completely lined with mesoderm.

**Pseudopodium:** a flowing projection of a cell or unicellular organism.

**Recessive:** the one of two alternative characters that is not evident in a heterozygous individual; opposite of *dominant*.

**Reflex arc:** neurons connecting a sensory receptor with an effector.

**Replication:** reproduction involving duplication; a doubling.

**Reproduction:** the process by which a species is maintained from generation to generation.

**Respiration:** in cells, the oxidation of foods; all steps involved in taking in oxygen and giving off carbon dioxide.

**Rhizoid:** a rootlike structure lacking vascular tissue.

**Rhizome:** an underground stem.

**Ribosomes:** microscopic granules in the cytoplasm, composed of RNA and protein; the site of protein synthesis.

**RNA:** ribonucleic acid, involved in protein synthesis.

**Root:** the plant organ whose primary functions are anchorage and absorption.

**Saprophytic:** type of nutrition in which the organism absorbs food through its walls.

**Schizocoel:** a coelom formed by a split within the mesoderm of the embryo.

**Sclerenchyma:** strengthening tissue in plants.

**Seed:** the structure in gymnosperms and angiosperms that consists of the embryo and associated tissues.

**Segregation, principle of:** two determiners for the same character cannot occur in the same gamete. (Mendel's first law.)

**Sensory:** responding to a stimulus.

**Sepal:** one of the units of the calyx of a flower.

**Sexual reproduction:** reproduction involving two individuals of opposite sex.

**Skull:** the skeleton of the head in vertebrates.

**Soma:** body cells, in contrast to germ cells or germ plasm.

**Sorus:** a group of sporangia.

**Species:** a group of similar organisms, the basic unit in classification; a group of interbreeding (or potentially interbreeding) individuals isolated reproductively from other such groups.

**Spermatogenesis:** the process of maturation of sperm cells.

**Sperm cell:** a male gamete.

**Sporangium:** a spore-producing structure.

**Spore:** a cell capable of developing independently into a new individual.

**Sporogenesis:** the production of spores; in the sporophyte this takes place by meiosis.

**Sporophyll:** scalelike structure or modified leaf bearing sporangia.

**Sporophyte:** spore-bearing generation in plants.

**Stamen:** the pollen-producing organ of the flower.

**Stele:** the vascular column of root or stem, including as its outer portion the pericycle.

**Stem:** the plant organ that functions primarily in support of the leaves and in conduction of liquids.

**Stomata:** openings in the leaf epidermis, regulated by paired guard cells.

**Strobilus:** a structure of the form of a cone, consisting of sporophylls attached to an axis; a cone.

**Substrate:** the substance acted upon by an enzyme.

**Substratum:** the material upon which an organism lives.

**Succession:** a natural sequence of changes in a biotic community leading to a climax community.

**Synapse:** the place where transmission from one neuron to another occurs.
**Synapsis:** lengthwise joining of homologous chromosomes during meiosis.
**Synecology:** the ecology of community relations.

**Taxonomy:** the science of classifying and naming plants and animals.
**Territorial behavior:** behavior associated with defense of an area by an animal.
**Testis:** the male gonad, producing spermatozoa.
**Thallus:** a plant body undifferentiated into root, stem, and leaves.
**Thorax:** the major division of the animal body next posterior to the head.
**Tissue:** a group of cells having the same function and structure.
**Tracheae:** air tubes.
**Transcription:** synthesis of a nucleotide sequence in messenger RNA complementary to the corresponding DNA sequence.
**Translation:** synthesis of protein through the action of messenger RNA.
**Transpiration:** controlled evaporation from leaves.
**Trophic level:** a group of organisms occupying the same level in a food chain.
**Tropism:** plant growth movement in response to directional stimulus.
**Turgor pressure:** externally directed pressure of plant cell membranes against their cell walls, resulting from higher internal osmotic pressure.

**Umbilical cord:** the connection between fetus and placenta in mammals.
**Uterus:** an enlarged portion of the female reproductive tract in which embryos are retained in ovoviparous and viviparous animals until birth.

**Vacuole:** a globule of liquid suspended in the cytoplasm.
**Vascular tissue:** general term applied to tissues that form the conducting channels in plants; xylem and phloem.
**Vector:** an organism that transmits a pathogenic organism from one host to another.
**Vegetative:** pertaining to nonreproductive functions.
**Vein:** a blood vessel carrying blood toward the heart.
**Ventral:** toward the lower side, away from the back.
**Ventricle:** a chamber of the heart from which blood leaves the heart.
**Villi:** minute fingerlike projections extending into the cavity of the intestine.
**Virus:** subcellular biological entity of protein and nucleic acid capable of reproducing only in living cells.
**Vitamin:** an essential dietary supplement, not used as a source of energy but required in enzyme systems.

**Xylem:** vascular tissue involved chiefly in conducting sap upward in a plant.

**Yolk sac:** the cavity containing the yolk of the embryo, opening into the embryonic gut; its covering is one of the embryonic membranes.

**Zoology:** the science of animal life.
**Zygote:** a fertilized egg.

# *Appendix E*

# SAMPLE FINAL EXAMINATIONS

Two kinds of examination questions are readily distinguishable, the essay type, which requires an organized discussion as an answer, and the objective type, which requires answers only in words, phrases, or simply check marks.

## AN ESSAY-TYPE EXAMINATION

Questions to be answered in essay form are typically quite general, rarely specifying any of the details expected of the student. They are not necessarily in interrogative form, so they are "questions" only in the broadest sense. They may use such directives as "discuss," "compare," or "explain." The following are sample essay-type final examination questions.

1. Discuss the role of carbohydrates in living organisms.
2. What is RNA and how does it function?
3. Compare blue-green algae and green algae.
4. How does the human kidney function?
5. What is territorial behavior?

In answering one of the above questions—or any essay question—the student should first plan his answer. An outline will help; it will not only result in better organization but actually save time, in the long run. It may be brief, and it need not be written down. Organization is important, however, for the two factors that produce a good essay are good organization and clear exposition. Of course you must know the essential facts; without them you have nothing to organize. But presenting such facts in a series of disconnected sentences does not produce a good essay. Aim for a logical sequence. If you have this, you can also develop a clear exposition from it if you adopt appropriate paragraphing and use complete sentences with adequate connectives. Your outline should provide the clue to breaking the essay into paragraphs.

In answering essay-type questions it is particularly important to budget your time with care. If you have a question as comprehensive as No. 1 above on a final examination with four other questions, you obviously cannot go into much detail. Nevertheless, certain aspects must be discussed, even briefly: carbohydrate manufacture by photosynthesis, with energy storage; condensation of carbohydrate food reserves; oxidation of carbohydrates in respiration, with energy release. Perhaps fermentation should be included, and the structural role of cellulose. If more time is available, expansion can include more details from the processes of photosynthesis and respiration. Answering the rather more circumscribed question on the functioning of the human kidney requires a rather natural outline: first the morphological background (gross structure of kidney and nephridium and their relations to circulatory system and urinary bladder), then the physiological account, which should have the major emphasis (primarily the functional differences of different parts of the nephridium but also including the functions of collecting channels and ureter). The amount of time available for the whole examination, together with differences in scope among the questions, should determine the way in which you budget your time.

While not either essay or objective in form, one common type of examination question is a request for definitions of terms. In defining a term, the principal thing to remember is that you must set limits around the concept defined in such a way that your definition will not serve for anything else. To do this effectively requires, first of all, placing the concept in a higher category that partially limits it. Thus, a mitochondrion, a lysosome, and a centrosome are all cytoplasmic organelles; mitosis, photosynthesis, and phosphorylation are all processes; Orthoptera, Hymenoptera, and Diptera are all orders of the Insecta. After selecting the category, distinguish the concept defined from others in the same category. Thus, a centrosome is a cytoplasmic organelle from which spindle fibers radiate during mitosis; Diptera are insects with but a single pair of wings.

# AN OBJECTIVE EXAMINATION

Answers to an objective examination require thorough understanding of vocabulary. Vocabulary is the key to any objective method of testing scientific knowledge. A full understanding of the technical language depends, however, upon knowledge of both facts and principles. Hence, both objective and essay-type questions test for facts. In objective examinations much more material can be covered in a given length of time. This is one reason objective examinations are widely used. Another reason is that papers may be graded more "objectively," therefore more justly, where there is a clear-cut answer to each question. And, of course, another reason is that many classes are so large that grading essay-type answers with fairness to all requires more time than is available to the instructional staff.

The following sample final examination does not necessarily correspond in length or content to an average final examination. It is designed primarily to suggest the various kinds of objective tests in common use in biology, and it contains a sample of each with the exception of one kind—an unlabeled drawing from a laboratory exercise, to which the student is expected to add the appropriate labels.

**Multiple Choice.** In each of the following sentences two or three alternative words or phrases are given. Choose the one that will make each sentence read correctly.

1. A group of organisms bound together by environmental factors constitutes
   a. a habitat
   b. a community
   c. a succession

2. An organism that manufactures its own food is
   a. autotrophic
   b. heterotrophic
   c. saprophytic

3. Oxidation taking place in the absence of oxygen is
   a. reducing
   b. aerobic
   c. anaerobic

4. Protein is digested by
   a. trypsin
   b. lipase
   c. amylase

5. The human ovum after ovulation first enters the
   a. ovary
   b. Fallopian tube
   c. uterus

6. Cytoplasmic organelles that are centers of respiratory enzyme activity are
   a. chromatids
   b. vacuoles
   c. mitochondria

7. Meiosis invariably involves
   a. one cell division
   b. two cell divisions
   c. three cell divisions

8. Darwin's theory of evolution is known as the theory of
   a. mutations
   b. acquired characteristics
   c. natural selection

9. The ocean as a habitat for organisms differs from land and fresh water in its greater
   a. stability
   b. variability

10. The oxygen given off in photosynthesis is derived from
    a. water
    b. PGAL
    c. carbon dioxide

11. The genetic code is expressed in
    a. enzyme sequence
    b. amino acid sequence

    c. nucleotide sequence
12. Calcium balance in the human body is regulated by the
    a. parathyroid glands
    b. islets of Langerhans
    c. thyroid glands
13. Glycolysis takes place in
    a. darkness
    b. light
    c. either darkness or light
14. The number of separate cavities of the human coelom is
    a. two
    b. three
    c. four
15. The cavity of the aqueous humor is
    a. behind the lens
    b. in front of the lens
    c. within the lens

**Matching.** In this type of question sets of related terms are given in two columns, one of which is numbered. The numbers are used to match the terms in that column with those terms in the other column with which they correspond. The following examples illustrate some of the possible variations.

*Part I. Mitosis.* Each of the events listed in the first column below characterizes one of the stages in column two. Place the number of the event in front of the name of the appropriate stage, separating numbers by commas if more than one event occurs during the same stage.

1. Formation of mitotic spindle      ____ Interphase
2. Formation of daughter nuclei      ____ Prophase
3. Replication of DNA      ____ Metaphase
4. Disintegration of nuclear membrane    ____ Anaphase
5. Movement of chromosomes toward poles ____ Telophase
6. Establishment of the cell plate
7. First appearance of chromosomes as distinct units

*Part II. Phyla.* Each of the characteristics or conditions listed in the first column is appropriate to one (and one only) of the phyla in column two. Place the number of the characteristic in front of the name of the phylum to which it applies.

1. Stinging cells      ____ Bryophyta
2. Hyphae      ____ Coelenterata
3. Without definite nuclei      ____ Annelida
4. Sporophyte parasitic on green gametophyte ____ Eumycophyta
5. Pharyngeal gill slits      ____ Chordata
6. Xylem present      ____ Arthropoda
7. Tube feet      ____ Cyanophyta
8. Metameric coelom      ____ Chlorophyta
9. Thallus plants that store starch    ____ Tracheophyta
10. Chitinous exoskeleton      ____ Echinodermata

**Completion.** In the following paragraphs certain essential terms have been omitted, their places being taken by numbers in parentheses. In the spaces provided, insert the appropriate terms.

The science dealing with all living things is called (1) _____, that subdivision dealing with plants being (2) _____ and that dealing with animals being (3) _____. Either of these may be subdivided into (4) _____, the study of function, and (5) _____, the study of structure.

*(The following statements apply to man.)*

The location of a muscle is given by naming the place of attachment of each end, the end moving more during contraction being the (6) _____ and the end moving less being the (7) _____. In the biceps muscle these two ends are attached respectively to the (8) _____ and the (9) _____. The action of the biceps is (10) _____ of the arm at the (11) _____, illustrating a (12) _____ class lever. The opposing action is accomplished by the (13) _____ muscle, operating as a (14) _____ class lever, the end with greater movement being attached to the (15) _____.

Oxygen, necessary in the recovery phase of muscular contraction, is carried to the muscles in loose combinations with (16) _____, a pigment contained in blood cells called (17) _____. The waste products of cellular oxidation are (18) _____ and (19) _____. These are carried in the blood to the heart, where they empty into the (20) _____ atrium, thence passing into the (21) _____. From the latter chamber blood flows to the lungs through the (22) _____. In the lungs the blood flows through microscopic vessels called (23) _____, which lie in the walls of the air sacs or (24) _____. Movement of gases between the cavities of the air sacs and the blood takes place by (25) _____.

**Completion of a Table.** In the following table of parts of a typical flower, the four whorls of structures in a complete flower are numbered from outside in, the outer whorl being Whorl 1. Complete the table.

| | Collective name | Name of individual part |
|---|---|---|
| Whorl 1 | | |
| Whorl 2 | | |
| Whorl 3 | | |
| Whorl 4 | | |

**True-False.** Place a T in front of each true statement, an F in front of each false one. Remember that if any part of a statement is false the whole statement must be considered false.

1. _____ A single cell may be as much as an inch in diameter.
2. _____ A eucaryotic cell is one that lacks double-membrane organelles.
3. _____ A cell wall is a typical constituent of animal cells.
4. _____ Higher osmotic pressure outside a plant cell causes plasmolysis.
5. _____ The peptide linkage involves the carboxyl group of one amino acid and the amino group of another.
6. _____ ATP is the major hydrogen acceptor in living cells.

7. \_\_\_\_ In asexual reproduction the progeny are derived from a single parent.
8. \_\_\_\_ A fruit is the ripened ovule.
9. \_\_\_\_ Yeasts carry on autotrophic nutrition.
10. \_\_\_\_ *Paramecium* reproduces by both sexual and asexual methods.
11. \_\_\_\_ Malaria is caused by a protozoan.
12. \_\_\_\_ In classification, orders are combined to form families and families to form genera.
13. \_\_\_\_ Basidia are borne on the gills of mushrooms.
14. \_\_\_\_ Cells in the leaves of a moss plant are diploid.
15. \_\_\_\_ The fern sporophyte contains vascular tissue.
16. \_\_\_\_ In flowering plants, the egg nucleus is but one of several haploid nuclei in the female gametophyte.
17. \_\_\_\_ Tropisms in plants are the result of differential rates of growth.
18. \_\_\_\_ Cell division occurs in all parts of higher plants.
19. \_\_\_\_ Monocots have more seed leaves than dicots.
20. \_\_\_\_ The excretory system of the earthworm is metameric.
21. \_\_\_\_ Complete metamorphosis of insects is development in which the hatched young are quite similar to the adults.
22. \_\_\_\_ The appendages of a crayfish are modified from a biramous type.
23. \_\_\_\_ All Chordata have a dorsal, tubular nerve cord.
24. \_\_\_\_ Independent assortment involves the separation of two alleles.
25. \_\_\_\_ A conditioned reflex may involve a minimum of three neurons.

**Arrangement in Sequence.** Trace the blood through the path it must follow by the *shortest route* in each of the following cases, naming all heart chambers, arteries, veins, and capillary systems in order.

a. In the frog, lung capillaries to mesonephric capillaries:
1. lung capillaries          6.
2.          7.
3.          8.
4.          9. mesonephric (renal) capillaries
5.

b. In man, inferior vena cava to common iliac artery:
1. inferior vena cava          6.
2.          7.
3.          8.
4.          9.
5.          10. common iliac artery

**Lists:**
1. List and characterize the major types of compounds in protoplasm.
2. Name the twelve cranial nerves of higher vertebrates and give their distributions.
3. List four developments that took place in human evolution.
4. Give the complete classification of one species of plant or animal, using six categories.

**Genetics:**

a. In guinea pigs black or pigmented hair (P) is dominant over white (p), the unpigmented condition; and rough hair (R) is dominant over smooth (r). A female with smooth white hair has a litter of five young, two of these black with rough hair and the other three white with smooth hair. Give genotypes of both parents and the five young:

Parents:               Male:          Female:
Progeny:

b. Four-o'clock plants with pink flowers are the heterozygous progeny of homozygous red (RR) and homozygous white (rr) parents. Give the expected phenotypic and genotypic ratios of progeny in the following three crosses:

| | | | |
|---|---|---|---|
| (1) | Pink | X | White |
| (2) | Pink | X | Red |
| (3) | Pink | X | Pink |

# ANSWERS

**Multiple Choice:** 1, b; 2, a; 3, c; 4, a; 5, b; 6, c; 7, b; 8, c; 9, a; 10, a; 11, c; 12, a; 13, c; 14, c; 15, b.

**Matching:**

*Part I. Mitosis.* 3, Interphase. 1, 4, 7, Prophase. (0) Metaphase. 5, Anaphase. 2, 6, Telophase.

*Part II. Phyla.* 4, Bryophyta. 1, Coelenterata. 8, Annelida. 2, Eumycophyta. 5, Chordata. 10, Arthropoda. 3, Cyanophyta. 9, Chlorophyta. 6, Tracheophyta. 7, Echinodermata.

**Completion:** 1. biology. 2. botany. 3. zoology. 4. physiology. 5. morphology. 6. insertion. 7. origin. 8. radius. 9. scapula. 10. flexing. 11. elbow. 12. third. 13. triceps. 14. first. 15. ulna. 16. hemoglobin. 17. erythrocytes. 18–19. carbon dioxide and water. 20. right. 21. right ventricle. 22. pulmonary arteries. 23. capillaries. 24. alveoli. 25. diffusion.

**Table:**

| Whorl 1: | calyx | sepal |
|---|---|---|
| Whorl 2: | corolla | petal |
| Whorl 3: | androecium | stamen |
| Whorl 4: | gynoecium | carpel |

**True-False:**

*True:* Nos. 1, 4, 5, 7, 10, 11, 13, 15, 16, 17, 20, 22, 23.

*False:* Nos. 2, 3, 6, 8, 9, 12, 14, 18, 19, 21, 24, 25.

**Arrangement in Sequence:**

a. 2. pulmonary veins. 3. left atrium. 4. ventricle. 5. conus arteriosus. 6. systemic arch. 7. dorsal aorta. 8. renal arteries.

b. 2. right atrium. 3. right ventricle. 4. pulmonary arteries. 5. lung capillaries. 6. pulmonary veins. 7. left atrium. 8. left ventricle. 9. aorta.

**Lists:**

1. See Chapter II.
2. See Chapter XI, especially Table 11.2.
3. See Chapter XVIII.
4. See Chapter XIX.

**Genetics:**

a. Parents:      Male:   PpRr       Female:   pprr

   Progeny:      2 young:   PpRr

                    3 young:   pprr

b. (1) Progeny: 1 pink (Rr) : 1 white (rr)

   (2) Progeny: 1 pink (Rr) : 1 red (RR)

   (3) Progeny: 1 red (RR) : 2 pink (Rr) : 1 white (rr)

# INDEX

# INDEX

365

Index 365

Imbibition, 38
Immunity, 302
Implantation, 235
Imprinting, 295
Impulse (nervous), 192–193
Incus, 230–231
Independent assortment, 240
Induced enzymes, 61
Induction (embryonic), 203
Inflorescences, 118
Inheritance
  acquired characters, 252, 263
  extrachromosomal, 252
Innate behavior, 285
Insecta, 159–162
Instinctive behavior, 288
Internode, 114
Interphase, 64–65
Intertidal zone, 275
Intestine, 182–216
Intrauterine device (IUD), 235
Invagination, 201
Iris, 184, 194, 229
Ischium, 181, 212
Islets of Langerhans, 227
Isogametes, 83, 95, 133, 136
Isolation and evolution, 263
Isolecithal, 200
Isoptera, 161–162
Isotonic solutions, 37
Isozymes, 41

Jacob, 60
Java man, 260–262
Jejunum, 216

Kalanchoe, 132
Karyokinesis, 64
Kelp, 137
Kidney, 225
Kineses, 287
Kinetochore, 64
Kinetosome, 33
Kingdoms, 72, 75
Kinins, 124
Kornberg, 57, 320
Krebs, Hans, 320
Krebs cycle, 43, 53

Labium, 167
Labrum, 167
Lactation, 227, 236
Lacteals, 219, 222
Lactic acid, 49, 190, 215
Lamarck, 1, 27, 260, 263, 316, 317
Lamellae, 33
Laminaria, 103, 137
Larynx, 182, 223
Lavoisier, Antoine, 318

Leaf
  arrangement, 114
  blade, 103
  structure, 115
Leakey, 260
Learning, 285, 290
Legume, 119
Lemma, 121
Lens, 184, 194
Lenticel, 115
Lepidoptera, 161–162
Leucocytes, 191, 220
Leucoplasts, 33
Liebig, Justus, 317, 318
Life
  characteristics, 4, 321
  origin, 6, 322
  physicochemical basis, 9
Life cycles
  brown algae, 136–137
  cell, 66–67
  fern, 139
  flowering plants, 140–142
  fungi, 135
  green algae, 136
  moss, 137–138
  pine, 139–140
Life zones, 283
Limnology, 275
Linkage, 249
Linnaeus, C., 267, 318
Lipases, 188, 218
Lipids, 19
Littoral organisms, 275
Liver, 182, 217, 225
Liver fluke, 150
Liverworts, 104–106, 137–138
Locus, 247
Loop of Henle, 225
Lophophore, 152
Lorenz, Konrad, 320
Lumbar vertebrae, 212
Lumbricus, 155–158
Lung, 223, 225
Lycopsida, 106, 138
Lymphatic system, 222
Lysis, 76
Lysosomes, 32, 66

Macronucleus, 87
Malaria, 87–88, 302
Malleus, 230–231
Malpighian tubule, 171
Malthus, Thomas, 310
Maltose, 19
Mammalia, 175
Mammary glands, 210, 236
Man
  development, 234–236
  morphology, 209–215

Nerve
  coordination, 192–193
  impulse, 229
  tissue, 95–96
Nervous system
  crayfish, 165
  earthworm, 158
  frog, 184
  grasshopper, 169
  honey bee, 171
  man, 227–229
  mussel, 154
  shark, 178
Nest building, 294
Neurohumor, 193, 229
Neurons, 192, 229
*Neurospora,* 101, 135
Ngandong man, 262
Niacin, 123
Niche, 278
Nicotin (vitamin B), 46, 219
Nirenberg, Arthur, 59, 320
Nitrogen equilibrium, 219
Node, 114
Nomenclature, 269–270
Notochord, 173
Nucellus, 139, 141
Nuclear body, 79
  granule, 67
  membrane, 34
Nucleic acids, 17, 23
Nucleoli, 34
Nucleoplasm, 34
Nucleus, 28, 34, 56
Nut, 119
Nutrient cycles, 279
Nutrition, 216, 219

*Obelia,* 135, 148–149, 196
Ocelli, 172
Oculomotor nerve, 179
Odonata, 161–162
*Oedogonium,* 103, 136
Olfactory nerve, 179
Onycophora, 158
Oocyte, 199
Oögenesis, 198–199
Oögonia, 136, 199
Operator gene, 60
Operculum, 175
Operon, 60
Optic nerve, 179, 184
Oral groove, 87
Orders (classification), 75, 267–269
Organ of Corti, 230
Organ systems, 4, 95
Organelles, 4, 28
Organic chemistry, 13
Organismal concept, 93
Organismic hypothesis, 322

Organizer, 203
Organogeny, 203
Organs, 4, 95
Ornithine cycle, 191
Orthogenesis, 264
Orthoptera, 159–161
*Oscillatoria,* 80
Osculum, 145
Osmosis, 37
Ossicles, auditory, 212, 230–231
Osteichthyes, 175
Ostia, 164
Ovary, 116, 197
Oviparous, 203
Ovipositor, 169
Ovoviviparous, 203
Ovulate cones, 109
Ovulation, 233
Ovules, 105, 109, 139, 141, 197, 233
Oxidation-reduction reactions, 42
Oxidative phosphorylation, 43, 53–54
Oxygen transport, 223
Oxyhemoglobin, 223–224

Paedogenesis, 197
Palea, 121
Paleontology, 253
Palolo worm, 289
Pancreas, 217, 218
Panicle, 118
Pantothenic acid, 219
*Paramecium,* 86–87, 288
Paramylum, 86
*Paranthropus,* 266
Parasite-host relationships, 79, 277
Parasympathetic system, 228
Parathyroid gland, 226
Parazoa, 143
Parenchyma, 95–96, 107, 116
Parthenogenesis, 197, 199
Parturition, 236
Pasteur, Louis, 81, 236, 319
Patella, 212
Pathogens, 203
Peck order, 294
Pectoral girdle, 177, 181, 212
Pelagic organisms, 275
Pellicle, 85
Pelvic girdle, 177, 212
Pelvis, 225
*Penicillium,* 100
Penis, 200
Pepsin, 218
Pepsinogen, 218
Peptide bond, 21
Perianth, 116
Pericardial cavity, 177, 209
Pericarp, 119
Pericycle, 112
Perigynous flowers, 117